BIOCHEMISTRY ILLUSTRATED

Dedication to the first edition
We would like to dedicate this book to our wives, Mollie and Shirley and our children, Alastair (a paediatrician), Julia (a Health Visitor) and Corinne (a budding biochemist), all of whom have been affected by, and sometimes afflicted by, our love for biochemistry.

Dedication to the second edition
In addition to the above, one of us as an indicator of progress, would like to add his four grandchildren, at least one of whom already finds the book interesting.

BIOCHEMISTRY ILLUSTRATED

AN ILLUSTRATED SUMMARY OF
THE SUBJECT FOR MEDICAL AND
OTHER STUDENTS OF BIOCHEMISTRY

PETER N. CAMPBELL
PROFESSOR OF BIOCHEMISTRY

ANTHONY D. SMITH
READER IN BIOCHEMISTRY

University College London Medical School, London

ILLUSTRATORS: **SUE HARRIS** and **JANE TEMPLEMAN**

THIRD EDITION

CHURCHILL LIVINGSTONE
EDINBURGH LONDON MADRID MELBOURNE NEW YORK
AND TOKYO 1994

CHURCHILL LIVINGSTONE
Medical Division of Longman Group UK Limited

Distributed in the United States of America by Churchill Livingstone Inc., 650 Avenue of the Americas, New York, N.Y. 10011, and by associated companies, branches and representatives throughout the world.

First edition 1982
Second Edition 1988
Third edition 1994
Reprinted 1994

ISBN 0-443-04573-9

British Library Cataloguing in Publication Data
A catalogue record for this book is available from the British Library.

Library of Congress Cataloging in Publication Data
Campbell, P. N. (Peter Nelson)
Biochemistry illustrated: an illustrated summary of the subject for medical and other students of biochemistry/Peter N. Campbell, Anthony D. Smith; illustrators, Sue Harris and Jane Templeman.
– 3rd ed.
p. cm.
Includes index
ISBN 0-443-04573-9
1. Biochemistry—Outlines, syllabli, etc. I. Smith, A.D. (Anthony Donald) II. Title.
[DNLM: 1. Biochemistry—Outlines.
QU 18 C189ba 1993]
QP518.3.C36 1993
574.19/2-dc20
DNLM/DLC
for Library of Congress 93-3365
CIP

The publisher's policy is to use **paper manufactured from sustainable forests**

Produced by Longman Singapore Publishers Pte Ltd
Printed in Singapore

PREFACE TO THE THIRD EDITION

Although this third edition is only about 15 pages of text longer than the second edition, we have reassessed every page and as a result have incorporated more than 150 new figures. Some of the more simplistic figures in the previous editions have been removed to allow us to incorporate developments at the growing points of biochemistry and molecular biology.

Although in producing the first edition we did not intend that the text should be a substitute for the standard textbooks we are aware that there is an increasing call for a slim text. There are three main reasons for this. First, the increase in our knowledge of biochemistry and molecular biology means that many of the standard texts are beyond the reach of many students in terms of both bulk and cost. Second is the demand of students and teachers that the amount of factual material in their courses should be reduced. We see this in the request of the General Medical Council in the UK for a core curriculum for medical students. Third, an increasing lack of understanding of English among students in many countries where formerly it was well taught. In producing a third edition we have taken account of these tendencies by increasing the amount of text so that the book might serve as a stand-alone textbook with suitable supplementation from lectures, tutorials and other reading.

Our illustrator, Sue Harris, was not available for this edition. We have been very fortunate in having the cooperation of our new illustrator, Jane Templeman, who has readily understood our requests and has produced many fine illustrations. We are indeed grateful to her.

Finally, it is a pleasure to thank all those friends throughout the world who have sent in their corrections and suggestions for improvements. We hope that with a new edition we will continue to receive such suggestions but hopefully not so many that concern our errors.

1993   P.N.C.
                                                           A.D.S.

PREFACE TO THE FIRST EDITION

The purpose of this book is to provide a survey of biochemistry in an easily assimilable form. We had in mind firstly students who might appreciate a succinct summary of some of the tenets on which a more advanced study of biochemistry is based. We hoped we might even encourage some students to develop their thirst for more extensive texts. Secondly we were aware of the needs of students in developing countries, who may not be too conversant with English. We thought it might be helpful to the students if the teachers could refer to clear diagrams in a modest book. Thirdly we wished to assist all those who, while not specializing in biochemistry, want to be aware of the major trends in the subject. In this respect we hope the book may prove useful to school teachers, physical scientists and medical doctors.

In making our choice of subject matter we have aimed at the student of medicine and the biochemistry student who is primarily interested in animals. We are aware that our choice has been arbitrary and we would be delighted to have suggestions concerning the subject matter and how we could improve the presentation.

We have of course worked within certain constraints. Firstly we have confined ourselves to about 200 pages, and secondly we have used only one colour in addition to black and white. As authors we consider we have been very fortunate in our collaborators: the artist, Sue Harris, worked closely with us, producing clear diagrams from our rough outlines; and the publishers have helped to keep us on sensible lines. In addition we thank Professor Maharani Chakravorty, Professor A. D. Patrick and Dr Ton So Har for their critical reading of the manuscript, and assisting us to ensure accuracy. We are more than grateful to all the numerous people and organizations who have given permission for the use of figures which we have redrawn from previous publications. We have a special debt to all our colleagues in the Courtauld Institute and elsewhere on whom we have called for expert advice on many occasions.

1982   P.N.C.
A.D.S.

CONTENTS

3

STRUCTURE AND FUNCTION OF ENZYMES

4

NUCLEIC ACIDS AND PROTEIN BIOSYNTHESIS

5 COENZYMES AND WATER-SOLUBLE VITAMINS

6 CARBOHYDRATE CHEMISTRY AND INTERCONVERSIONS OF MONOSACCHARIDES

7 NITROGEN METABOLISM

8 CARBOHYDRATE AND FAT METABOLISM

A. Oxidative catabolism

C. The absorptive state — synthesis of energy storage compounds

D. Phospholipids, other lipid substances and complex carbohydrates

9   MEMBRANE STRUCTURE AND FUNCTION

INDEX

A NOTE ON THE LAYOUT

The left hand column contains the section headings and a very brief summary of the subject matter. Further details are contained in the right hand column which is also where the illustrations are presented.

Each page is numbered so that it seemed unnecessary to number separately the illustrations. Where more than one illustration is presented they are lettered A, B, C, etc. It should therefore be easy for a lecturer to refer the students to a particular illustration.

At the first mention of an important word it is presented in italics. The word will also appear in the index together with other entries.

Rather than acknowledge each figure separately we have grouped the acknowledgements together. The division of the text into frames does not in any way imply that the book can be used as a teaching programme. It is intended that it be used in conjunction with more complete textbooks and a course of lectures.

The arrangement of carbohydrate and fat metabolism is unusual, but is based on experience of teaching medical students over a number of years. The division of pathways involved in the synthesis and degradation of energy-storage compounds into the fed and fasting states of metabolism has been found to give students a lasting understanding of the functions of these pathways. It also leads to a ready analysis of many clinical metabolic problems. Structural lipids and carbohydrates are described under the subject of membranes, to which they are functionally related.

Whilst some of the diagrams may seem to contain more information than appears necessary in a short summary (e.g. detailed structures of active sites), we felt that such additional detail does not in any way obscure the essential concept being conveyed, and gives an authentic basis for a more precise examination of the concept by students who wish to think more deeply about the subject.

ACKNOWLEDGEMENTS

The authors are pleased to acknowledge the source of those illustrations in the book which have been based on illustrations published elsewhere and to acknowledge the assistance of those who have helped in the preparation of various illustrations. Our acknowledgements have been grouped as follows:

a. The authors and publishers of books and original reports in journals. This list may also be of value to those who wish to extend their reading.
b. Those persons and companies who have provided material from their own sources.

In all cases the numbers in [ ] refer to the number of the illustration in the present book. The other numbers refer to the original source.

a. References to books and journals

1. General textbooks
1.1. *Basic Biochemistry for Medical Students* (Campbell, P.N. & Kilby, B.A., eds) Academic Press, London.
Apart from the editors, the authors were J.B.C. Findlay, H. Hassall, R.P. Hullin, A.J. Kenny and J.H. Parish.
Figs 2.8 [11B], 2.10 [12A], 2.15 [11A], 4.10 [20A], 4.16 [14A], 4.18 [15A], 5.24 [75A], 12.3 [93C], 12.12 [100B], 12.16 [103B], 12.23 [105A], 12.25 [106B], 12.27 [107A]. Tables 2.8 [15B], 4.2 [15C].

1.2. *Biochemistry: A Functional Approach* 3rd edn (McGilvery, R.W. & Goldstein, G.W.)
W.B Saunders, Philadelphia.
Fig. 3.12 [22B].

1.3. *Biochemistry with Clinical Correlations* 2nd edn (Devlin, T.M., ed.)
J. Wiley, New York.
Figs 3.6 [74A], 6.39 [132C], 6.50 [189A], 6.51 [189B].

1.4. *Biochemistry* (Stryer, L.)
W.H. Freeman, Oxford.
2nd edn, Figs 4.2 [25A], 4.3 [25B], 4.6 [25C], 4.20 [24B], 30.21 [97B].
3rd edn, Figs 2.44, 2.45 [19A], Fig 3.16 [12B], 29.24 [110A], 29.29 [123A].

1.5. *Principles of Biochemistry* 6th edn (White, A., Handler, P., Smith, E.L., Hill, R.L. & Lehman, I.R.)
McGraw-Hill, New York.
Figs 36.1, 36.2, 36.5, 36.6 [50A, B, C, D], 36.4, 36.9 [51A, B].

1.6. *Molecular Biology* (Freifelder, D.)
Jones and Bartlett, Boston.
Figs 4.32 [87A], 21.27 [98B].

1.7. *Biochemistry for the Medical Sciences* (Newsholme, E.A. & Leech, A.R.)
Wiley, Chichester.
Figs 7.9 [176A], 11.5 [201B].

1.8. *Biochemistry* (Mathews C.K. & van Holde K.E.)
Benjamin/Cummings, Redwood City, CA.
Figs 7.12 [23B], 11.1 [58B], 28.35 [117A], 28.36 [117B].

1.9. *Essential Immunology* 6th edn (Roitt, I.M.)
Blackwell Scientific, London.
Fig 7.10 [49A].

1.10 *Gray's Anatomy* 37th edn (Williams, P.L., Warwick, R., Dyson, M. & Bannister, L.H., eds)
Churchill Livingstone, Edinburgh.
Fig. 1.75B[146B].

1.11. *Harper's Biochemistry* 21st edn (Murray, R.K., Cranner, D.K., Mayes P.A. & Rodwell, V.W., eds)
Prentice-Hall, New York.
Fig. 55.2 [31A].

2. Books on special topics

2.1. *Open University Course Book* S322 Units 1–2 (1977)
Open University Press, Milton Keynes.
Fig. 3, p. 14 [18A, B, C, D].

2.2. *Cells and Organelles* 2nd edn (Novikoff, A.B. & Holtzmann, E.)
Holt, Rinehart & Winston, New York.
Figs. 1.23 [5A], 1.25 [6B].

2.3. *Structure and Action of Proteins* (Dickerson, R.E. & Geis, I.)
Benjamin, New York.
p. 47 [22A], 56 [23A].

2.4. *Enzyme Structure and Mechanism* 2nd edn (Fersht, A.)
W.H. Freeman, Oxford.
Figs 1.12 [62A, B], 1.13[65A], 10.3 [66B], 12.13 [61B].

2.5. *Chance and Necessity* (Monod, J., Trans. by Wainhouse, A.)
A.A. Knopf, New York, and William Collins, London.
p. 47, Fig. 4 [109A].

2.6. *The Structure and Function of Animal Cell Components* (Campbell, P.N.)
Pergamon Press, Oxford.
Figs 2.2 [190], 5.2 [161B].

2.7. *Advancing Chemistry* (Lewis, M. & Waller, G.)
Oxford University Press, Oxford.
p. 311 [184].

2.8. *Supplement to DNA Replication* (Kornberg, A.)
W.H. Freeman, San Francisco.
Frontispiece [95B].

2.9. *Molecular Basis of Antibiotic Action* 1st edn (Gale, E.G., Cundliffe, E., Reynolds, P.E., Richmond, M.E. & Waring, M.J. eds)
Wiley Interscience, New York.
Cundliffe, E., p. 278 [99B]
Waring, M.J., p. 173 [107B]

2.10. *Structure of Mitochondria* (Munn, E.A. ed.)
Academic Press, London.
Kroger, A. & Klingenberg, M., p. 282 [183B].

2.11. *A Guided Tour of the Living Cell* (Christian de Duve)
Scientific American Library.
Illustration © 1982 Neil Hardy, p. 272 [113A], p. 344 [45A].

2.12. *Molecular Biology of the Cell* (Alberts, B., Bray, D., Lewis, J., Raff, M., Roberts, K. & Watson, J.D., eds) Garland, New York.
1st edn Figs 8.24 [92B], 8.59 [2C].
2nd edn Figs 11.35 [279B], 11.47 [280A].

2.13. *Immunology* (Eisen, H.N.)
Harper & Row, New York.
p. 132 [43A].

2.14. *Principles of Gene Manipulation* 3rd edn (Old, R.W. & Primrose, S.B.)
Blackwell Scientific, Oxford.
Figs 1.3, 1.4 [124A, B].

2.15. *From Cells to Atoms* (Rees, A.R. & Steinberg, M.J.E.)
Blackwell Scientific, Oxford.
Figs 10.1 [52A], 11.2 [38A], 26.1 [92A], 40.2 [44A].

2.16. *Recombinant DNA. A Short Course* (Watson, J.D., Tooze, J. & Kurtz, D.T.)
Scientific American Books.
Figs 5-3 [125], 5-4 [126].

2.17. *Immunology* 1st edn (Roitt, I.M., Brostoff, J. & Male, D.K.)
Churchill Livingstone, Edinburgh, and Gower, London
Figs 5.15 [42B], 7.8 [44B].

2.18. *Separation of Plasma Proteins* (Curling, J.M., ed.)
Pharmacia Fine Chemicals AB, Uppsala.
Fig. 39 [33B].

2.19. *Albumin, An Overview and Bibliography*
Miles Laboratories, IN.
Physiological transport functions of albumin [35A].

2.20. *The Ultrastructural Anatomy of the Cell* (Allen, T.D.)
Cancer Research Campaign, London.
[2B], [3A, B, C].

2.21. *Biochemical Messengers* (Hardie, D.G.)
Chapman and Hall, London.
Figs 6.6 [73B], 8.43 [69A]

2.22. *Molecular Biology and Biotechnology* (Smith, C.A. & Wood, E.J. eds)
Chapman and Hall, London.
Figs 4.4 [104A], 4.12 [104B], 5.8 [123B], 7.8 & 7.19 [111A].

2.23. *Gene Regulation* (Latchman, D.)
Unwin Hyman, London.
Figs 7.9 & 7.10 [110B], 7.13 [111B].

2.24. *The New Genetics and Clinical Practice* 3rd edn (Weatherall, D.J.)
Oxford University Press, Oxford.
Figs 46 [128B], 79 [128A].

2.25. *Proteins* (Creighton, T.E.)
W.H. Freeman, New York.
Fig. 5-6 [21A].

2.26. *New Trends in Biological Chemistry* (Ozawa, T. ed.)
Japan Scientific Societies Press, Tokyo, and Springer-Verlag, Berlin.
Steverding, D. & Kadenbach, B., p. 158, Fig. 1 [187A].

2.27. *Molecular Biology of Oncogenes and Cell Control Mechanisms* (Parker, P.J. & Katan, M., eds)
Ellis Horwood, Chichester.
p. 85, Fig. 5 [267B], p. 89, Fig. 7 [268A], p. 90, Fig. 8 [268B].

3. Reviews

3.1. *Companion to Biochemistry* (Bull, A.T., Lagnado, J.R., Thomas, J.O. & Tipton, K.F., eds)
Longman, London.
Campbell, P.N. (1979) 2, Fig. 8.1 [112A].

3.2. *FEBS Symposium* Vol. 53 (Rapoport, S. & Scherve, T., eds)
Pergamon Press, Oxford.
Grant, M.E., p. 29–41 [113B].

3.3. *The Plasma Proteins* (Putnam, F.W. ed.)
Academic Press, London.
Putnam, F.W., vol. III, p. 14 [41A].

3.4. *The Enzymes* 3rd edn (Boyer, P.D., ed.)
Academic Press, London.
Dickerson, R.E. & Timkovich, R., vol. XI, p. 441, Fig. 8 [63B].

3.5. *Essays in Biochemistry* (Campbell, P.N., Greville, G.D. & Dickens, F., eds)
Academic Press, London.
Hales, C.N. (1967) 3, p. 75, Fig. 1 [177B].
Grant, P.T. & Combs, T.L. (1970) 6, p. 76, Fig. 3 [82A].
Williamson, A.R. (1982) 18, p. 24, Fig. 13 [122B].
Byard, E.H. & Lange, B.M.H. (1991) 26, p. 16, Fig. 2 [282B].
Sekiguchi, K., Maceda, T. & Titani, K. (1991) 26, p. 41, Fig. 2 [284A], p. 40, Fig. 1 [284B].

3.6. *Current Topics in Cell Regulation* (Horecker, B.L. & Stadtman, E.R., eds)
Academic Press, London.
Masters, C.J. (1977) 12, p. 77, Fig. 2 [66A].

3.7. *Trends in Biochemical Sciences*
Elsevier/North Holland, Amsterdam.
Benesch, R. (1978) 3, N 126 [26B].
Huber, M. (1979) 4, p. 271, Fig. 7 [41C].
Spiegel, A.M., Backlund, P.S. Jr, Butrynski, J.E., Jones, T.L.Z. & Simonds, W.F. (1991) 16, p. 339, Fig. 1 [266A].

3.8. *Seminars in Hematology*
Grune and Stratton, New York.
Rachmilewitz, E.A. (1974) 11, p. 453, Fig. 5 [27B].

3.9. *Annual Reviews of Medicine*
Annual Reviews, CA.
Stamatoyannopoulos, G., Bellingham, A.J., Lenfant, C. & Finch, C.A. (1971) *22*, p. 224 Fig. 1 [28A].

3.10. *Annual Reviews of Biochemistry*
Annual Reviews, CA.
Bennett, V. (1985) *54*, p. 283, Fig. 1 [279B].
McIntosh, J.R. & Snyder, J.A. (1976) *45*, p. 706, Fig. 1 [282A].
Klee, C.B., Crouch, T.H. & Richman, P.G. (1980) *49*, p. 496, Fig. 1 [269A].
Ferguson, M.A.J. & Williams, A.F. (1988) *57*, p. 292, Fig. 1 [266B] modified.
Yarden, Y. & Ullrich, A. (1988) *57*, p. 446, Fig. 2 [272A].
Strynadka, N.C.J. & James, M.N.G. (1989) *58*, p. 962, Fig. 1 [269B].
Edelman, G.M. & Crossin, K.L. (1991) *60*, pp. 158, 159, Fig. 1 [285A].
Kaziro, Y., Itoh, H., Kozasa, T., Nakafuku, M. & Satoh, T. (1991) *60*, p. 361, Fig. 5 [265A].
Dohlman, H.K., Thorner, J., Caron, M.G. & Lefkowitz, R.J. (1991) *60*, p. 662, Fig. 1a [259A].

3.11. *Scientific American*
Grobstein, C. (1977) July, p. 30 [118B].
Brown, M.S. & Goldstein, J.L. (1984) Nov., p. 55 [230], p. 56 [258A].
Rothman, J.E. (1985) Sept., p. 86 [253].
Dunant, Y. & Israel, M (1985) April, p. 42 [263A, B, C].
Lodish, H.F. & Dautry-Varsat, A. (1984) May, p. 51 [258A].
Gallo, R.C. (1987) Jan., p. 46 [98C].

3.12. *Biochimica et Biophysica Acta*
Elsevier/North Holland, Amsterdam.
Lotan, R. & Nicolson, G.L. (1979) *559*, p. 239 [257A].
Kagawa, Y. (1978) *505*, p. 47 [278A, B].
Small, D.M., Penkett, S.A. & Chapman, D. (1969) *176*, p. 178, Fig. 7 [228B].

3.13. *Biomedicine*
Springer International, Berlin.
Maclouf, J., Sors, H. & Rigaud, M. (1977) *26*, p. 362 [247A].

3.14. *Haemoglobin and Red Cell Structure and Function* (Brewer, G.J., ed.)
Plenum Press, London (1972).
Brenna, O., Luzzana, M., Pace, M., Perrella, M., Rossi, F., Rossi, F., Rossi-Bernardi, L. & Roughton, F.J.W., p.20, Fig. 1 [26C].

3.15. *Advances in Protein Chemistry*
Academic Press, New York (1981).
Richardson, J.S. *34*, p. 254, Fig. 71 [53A], p. 262, Fig. 73, p. 263, Fig. 74, p. 266, Fig. 77 [52A].

3.16. *Advances in Enzyme Regulation*
Pergamon Press, Oxford.
Saggerson, D., Ghadiminejad, I. & Awan, M. (1992) *32*, p. 286, Fig. 1 [204A].

3.17. *Biochemistry*
American Chemical Society, Washington, DC.
Huber, R. & Carrell, R.W. (1989) *28*, modified from p. 8960, Fig. 2 [64A].
Stroud, R.M., McCarthy, M.P. & Shuster, M. (1990) *29*, p. 11013, Fig. 3 [259B].

3.18. *Cell*
MIT Press, Cambridge, MA.
Cantley, L.C., Auger, K.R., Carpenter, C., Duckworth, B., Graziano, A. Kapeller, R. & Soltoff, S. (1991) *64*, p. 282, Fig. 1 [272B], p. 285, Fig. 4 [274A].

3.19. *Bioessays*
The Company of Biologists Limited, Cambridge.
Dustin M.L. (1990) *12*, p. 422, Fig. 1 [285A].

3.20. *Journal of Cell Science*
The Company of Biologist Limited, Cambridge.
Grinnell, F. (1992) *101*, p. 3, Fig. 1 [285C].

3.21 *Journal of Structural Biology*
Academic Press, Orlando.
Steinert, P.M. (1991) *107*, p. 187, Fig. 10 [283B].

4. Papers in journals

4.1. *Journal of Molecular Biology*
Academic Press, London.

Josephs, R., Jarosch, H.S. & Edelstein, S.J. (1976) *102*, p. 409, Fig. 6d [27A].

Valentine, R.C. & Green, N.M. (1967) *27*, p. 615 [43B].

Sigler, P.B., Blow, D.M., Matthews, B.W. & Henderson, R. (1968) *35*, p. 143, Fig. 6 [61A].

Rich, A. (1961) *3*, p. 483, Fig. 2 [38B].

4.2. *Biochemical Education*
International Union of Biochemistry and Pergamon Press, Oxford.

Hall, L. & Campbell, P. N. (1979) *7*, p. 57 [119, 120, 121].

Henderson, J.F. (1979) *7*, p. 52, Fig. 2 [137B].

Smith, I. (1980) *8*, p. 1 [40A].

4.3. *Science*
American Association for the Advancement of Science.

Britten, R.J. & Kohne, D.E. (1968) *161*, p. 530, Figs 1, 2 [88B, C].

Baulieu, E.-E. (1989) *245*, p. 1352, Fig. 2 [111C].

4.4. *Cell*
MIT Press, Cambridge, MA.

Lai, E.C., Stein et al (1979) *18*, p. 834, Fig. 6 [122A].

4.5. *Proceedings of the National Academy of Sciences, USA*
The National Academy of Sciences, Washington, DC.

Palade, G.E. (1964) *52*, p. 617, Fig. 2 [183B].

Siverton, E.L. (1977) *74*, p. 5142, Fig. 3 [41B].

4.6. *Philosophical Transactions of the Royal Society B*
The Royal Society, London.

Evans, P.R., Farrants, G.W. & Hudson, P.J. (1981) *293*, p. 53, Fig. 2B [68B].

4.7. *Journal of Cell Biology*
The Rockefeller University Press, New York.

Fernandez-Moran, H., Oda, T., Blair, P.V. & Green, DE. (1964) *22*, p. 73, Figs 6, 7 [187B].

Alexander, C.A., Hamilton, R.L. & Havel, R.J. (1976) *69*, p. 260, Fig. 14 [221B].

Osborn, M., Webster, R.E. & Weber, K. (1978) *77*, R.29 [283A].

4.8 *Nature, London*
Macmillan, London.

Arnone, A. (1972) *237*, p. 148 [24B].

Williams, A.F. (1984) *308*, p. 12 [46B].

Poorman, R.A. et al (1984) *309*, p. 468 [68A].

Ungewickell, E. & Branton, D. (1981) *289*, p. 420, Fig. 3 [276B].

Mishina, M. et al. (1985) *313*, p. 364, Fig. 1 [261A], Fig. 3, Table 1 [262A].

Barford, D. & Johnson, L.N. (1989) *340*, p. 609, Fig. 1 [199B].

4.9. *Biochemical Journal*
The Biochemical Society.

Andrews, P. (1964) *91*, p. 222 [30C].

4.10. *Journal of Biological Chemistry*
American Society of Biological Chemists.

Rosenberg, L., Hellmann, W. & Kleinschmidt, A.K. (1975) *250*, p. 1877, Fig. 1 [147A].

Trumpower, B.L. (1990) *265*, p. 11410, Fig. 1 [185B].

4.11. *Immunology Today*
Elsevier North Holland, Amsterdam.

Brodsky, F.M. (1984) *5*, p. 350, Fig. 4 [276A].

4.12. *Médecine Sciences*
CDR Centrale des Revues/John Libby Eurotext, Montrouge,

Hue, L. & Rider, M.H. *3*, p. 569, Fig. 2 [177A].

4.13. *Cell Motility and the Cytoskeleton*
Alan R. Liss, New York.

Lawson, D. (1987) *7*, p. 371, Fig. 2 [280A].

b. Personal acknowledgements

We wish to thank the following for detailed help with the diagrams indicated, or for allowing us to use diagrams already prepared:

5.1. Pharmacia Fine Chemicals AB, Box 175, S-7514, Uppsala 1, Sweden [30A, B; 33A].

5.2. Boehringer Mannheim GmbH, P O Box 310120, 6800 Mannheim 31, FRG [76A, B].

5.3. Dr Alberto N. Dutra, Dept of Histopathology, University College and Middlesex School of Medicine, and Dr Elinor M. Steen, lately of The Middlesex Hospital Medical School [4A].

5.4. Dr A.L. Miller, Dept of Chemical Pathology, University College and Middlesex School of Medicine and Miss Laura Worsley, lately of The Middlesex Hospital Medical School [34].

5.5. Dr G.L. Mills, lately of The Middlesex Hospital Medical School [231C].

5.6. Professor J.L.H. O'Riordan, Dept of Medicine, University College and Middlesex School of Medicine, and Dr L.M. Sandler, lately of The Middlesex Hospital Medical School [245B].

5.7. Bio-Rad Laboratories, Watford, Herts [32B].

5.8. Dr. B.R.F. Pearce, Lab. Mol. Biol., Cambridge [276B].

5.9. Professor R. Carrell, Dept of Haematology, University of Cambridge [35B, 36B, C, D].

5.10. Dr A. Moshtaghfard, lately of The Middlesex Hospital Medical School [31B].

5.11. Professor Janet Thornton and Dr. Marketa Svelebil, Dept of Biochemistry and Molecular Biology, University College London, for figures generated by MOLSCRIPT [64A].

5.12. Dr W.J. Whelan, University of Miami School of Medicine [214B], [215A].

5.13. Dr Graham Warren, ICRF Laboratories, London [275A].

5.14. Professor R.P. Ekins, University College and Middlesex School of Medicine, for comments on the section on free energy [70A].

5.15. Dr D.F. Steiner, Howard Hughes Medical Institute, University of Chicago [40A], [82A].

5.16. Professor K.F. Tipton, Dept of Biochemistry, Trinity College, Dublin [78C].

5.17. Professor K.R. Harrap, Cancer Research Campaign Laboratories, Sutton [96A], [108].

5.18. Dr J. Honour, Dept of Chemical Pathology, University College and Middlesex School of Medicine, London [240–242].

5.19. Professor J.S. Hyams, Dept of Biology, University College London [91A].

We also thank Dr A.-H. Etemadi, Laboratoire de Biochimie et de Biologie Moleculaire, Faculté des Sciences de l'Université Paris VI, Dr M.A. Rosemeyer and Professor E.D. Saggerson, Dept of Biochemistry and Molecular Biology, University College London and Dr F. Vella, Dept of Biochemistry, University of Saskatchewan, Saskatoon, for their comments on a variety of topics during the preparation of this edition.

KEY TO REFERENCES

Fig. No.	Ref. No.	Fig. No.	Ref. No.	Fig. No.	Ref. No.	Fig. No.	Ref. No.
2B	2.20	43A	2.13	104A,B	2.22	201B	1.7
2C	2.12	43B	4.1	105A	1.1	204A	3.16
3A–C	2.20	44A	2.15	106B	1.1	214B	5.12
4A	5.3	44B	2.17	107A	1.1	215A	5.12
5A	2.2	45A	2.11	107B	2.9	221B	4.7
6B	2.2	46B	4.8	108	5.17	228B	3.12
11A,B	1.1	49A	1.9	109A	2.5	231C	5.5
12A	1.1	50A–D	1.5	110A	1.4	240–2	5.18
12B	1.4	51A,B	1.5	110B	2.23	245B	5.6
14A	1.1	52A	2.15, 3.15	111A	2.22	247A	3.13
15A–C	1.1	53A	3.15	111B	2.23	253	3.11
18A–D	2.1	58B	1.8	111C	4.3	257A	3.12
19A	1.4	61A	4.1	112A	3.1	258A	3.11
20A	1.1	61B	2.4	113A	2.11	259A	3.10
21A	2.25	62A,B	2.4	113B	3.2	259B	3.17
22A	2.3	63B	3.4	117A,B	1.8	261A	4.8
22B	1.2	64A	3.17, 5.11	118B	3.11	262A	4.8
23A	2.3	65A	2.4	119	4.2	263A–C	3.11
23B	1.8	66A	3.6	120	4.2	265A	3.10
24B	1.4, 4.8	66B	2.4	121	4.2	266A	3.7
25A–C	1.4	68A	4.8	122A	4.4	266B	3.10
26B	3.7	68B	4.6	122B	3.5	267B	2.27
26C	3.14	69A	2.21	123A	1.4	268A,B	2.27
27A	4.1	70A	5.14	123B	2.22	269A,B	3.10
27B	3.8	73B	2.21	124A,B	2.14	272A	3.10
28A	3.9	74A	1.3	125	2.16	272B	3.18
30A,B	5.1	75A	1.1	126	2.16	274A	3.18
30C	4.9	76A,B	5.2	128A,B	2.24	275A	5.13
31A	1.11	78C	5.16	132C	1.3	276A	4.11
31B	5.10	82A	3.5, 5.15	137B	4.2	276B	4.8, 5.8
32B	5.7	87A	1.6	146B	1.10	227A,B	3.12
33A	5.1	88B,C	4.3	147A	4.10	279B	3.10
33B	2.18	91A	5.19	161B	2.6	280A	4.13
34A	5.4	92A	2.15	176A	1.7	280B	2.12
35A	2.19	92B	2.12	177A	4.12	281A	2.12
35B	5.9	93C	1.1	177B	3.5	282A	3.10
36A–D	5.9	95B	2.8	183B	2.10, 4.5	282B	3.5
38A	2.15	96A	5.17	184	2.7	283A	4.7
38B	4.1	97B	1.4	185B	4.10	283B	3.21
40A	4.2, 5.15	98B	1.6	187A	2.26	284A,B	3.5
41A	3.3	98C	3.11	187B	4.7	285A	3.10
41B	4.5	99B	2.9	189A,B	1.3	285B	3.19
41C	3.7	100B	1.1	190	2.6	285C	3.20
42B	2.17	103B	1.1	199B	4.8		

1

CELLULAR BASIS OF BIOCHEMISTRY

Living cells may contain a variety of organelles, including the nucleus, mitochondria, a network of membranes known as the endoplasmic reticulum, lysosomes and protein filaments such as comprise the microtubules and microfilaments that form the cytoskeleton. Cells can be disrupted in such a way as to preserve the function of these organelles. They can then be studied in isolation, and their content of different pathways and enzymes determined.

A

Types of living cells

Living cells may be subdivided into two groups:

Prokaryotes, e.g. bacteria
Eukaryotes, e.g. animals, plants, fungi, protozoa.

Note absence of mitochondria/chloroplasts and other organelles in bacteria. Chloroplasts are confined to plants.

Animal Cell

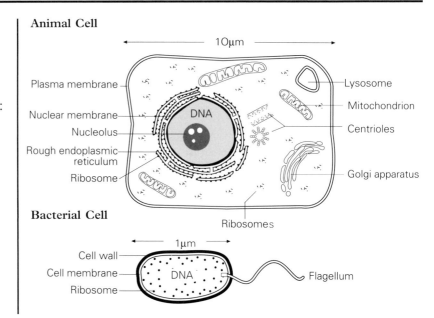

Bacterial Cell

B

The organelles of a typical animal cell

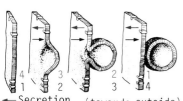

← Secretion   (towards outside)
→ Phagocytosis (towards inside)

Plasma membrane
This refers to the outer membrane of the cell in contact with the extracellular fluid. It has a particular role in secretion.
Vesicles from the cytoplasm fuse with the plasma membrane, which then ruptures to release the contents (exocytosis). Endocytosis involves the uptake of substances by invagination of the plasma membrane.

Cytoplasm
Strictly, this consists of all the components of a cell apart from the nucleus. It is often used, however, to mean all the components apart from the nucleus and the mitochondria.

Cytosol
This is that part of a cell in which we cannot readily detect organized components (organelles). It is not the same as the soluble supernatant obtained by disrupting a cell and centrifuging out all the particulate components (also often referred to as the cytosol). Such a soluble fraction will contain various soluble components extracted from the various organelles.

C

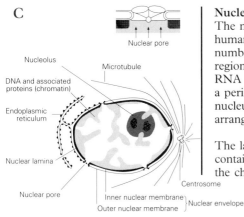

Nucleus
The nucleus contains the DNA organized into separate chromosomes. In a human diploid cell there are 23 pairs of chromosomes giving a diploid number of 46, with half as many in a haploid cell. The nucleolus is a region within the nucleus in which the genes for three of the four ribosomal RNA molecules are located. The nuclear membrane is double-layered with a perinuclear space. Transfer of substances between the cytoplasm and nucleus is through the nuclear pore, which is a complex of proteins arranged in an octagonal array with a central hole.

The lamina is a structure that surrounds the inner nuclear membrane and contains three proteins, lamins A, B and C. The lamina interconnects with the chromosomes.

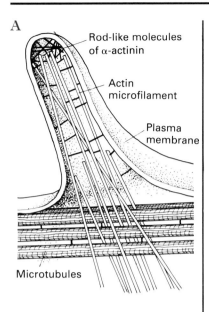

The Cytoskeleton

The high resolution that has been achieved with the electron microscope has revealed that there is a structural lattice even within the cytosol. This lattice is generally composed of three types of filaments, microfilaments based on the protein actin (see p. 280), microtubules based on the protein tubulin (see p. 282), and a heterogeneous group called intermediate filaments (see p. 283). Cell surface extensions are commonly supported by a core of actin microfilaments to produce an increase in the surface area as shown in the diagram of the lining of the small intestine.

Lysosomes

Lysosomes are membrane-bound vesicles containing a wide range of hydrolytic enzymes. They are central components in the *intracellular digestive system* so that, for example, substances and components of the cell that are to be degraded form *phagosomes* which fuse with the *lysosomes* to become *secondary lysosomes*.

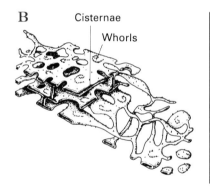

The endoplasmic reticulum

The endoplasmic reticulum is a network of membranes throughout the matrix of the cytoplasm. These membranes form a complex arrangement of connecting vesicles and tubules or large flattened sacs. The membranes run parallel to each other creating channels which are called cisternae. Large areas of membrane are thus created and it is estimated that the surface area of the endoplasmic reticulum in 1 ml of cytoplasm is 11 m^2. The surface may bear ribosomes (rough-surfaced endoplasmic reticulum) or may not (smooth-surfaced endoplasmic reticulum). The ribosomes are often characteristically arranged as 'whorls'.

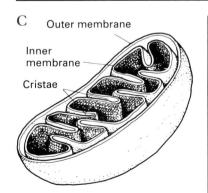

Mitochondria

Mitochondria are unique organelles within the cytoplasm, lacking any direct structural relationship with other organelles and containing their own DNA. Besides producing energy, they also help to control the level of calcium in the cytoplasm.

Mitochondrial membranes are about 6.5 nm thick with the inner membrane folded to form cristae. The inner surfaces of the cristae are closely packed with 8.5 nm particles, the sites of oxidative phosphorylation (see p. 187).

A

Types of biochemical preparations

Living cells may be studied at various levels of organization.

In assessing the physiological significance of a biochemical finding there is usually no substitute for the so-called intact animal. It is, however, difficult to control the various parameters in the whole body and so biochemists resort to other systems such as the incubation of single cells in carefully defined media. Organ explants (tiny pieces of tissue) have an advantage in that they can be effectively aerated with oxygen. The perfusion of intact organs represents a half-way stage.

Single cell, eukaryotic cells in culture
Scanning electron micrograph of a mouse peritoneal macrophage.

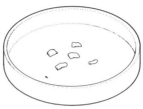

Organ explants, and slices of tissue

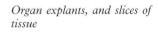

Atrial cannula — Aortic cannula

Left atrium

Right ventricle

Left ventricle

Intact organs
Perfusion: it is often convenient to study the metabolism of an intact organ under controlled conditions, e.g. a heart or liver.

B

Subcellular fractionation

Cells can be disrupted and the constituents separated by differential centrifugation.

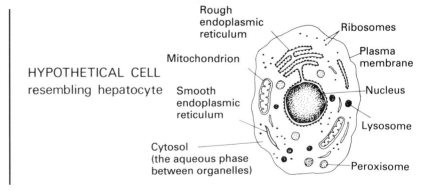

HYPOTHETICAL CELL resembling hepatocyte

Rough endoplasmic reticulum

Ribosomes

Plasma membrane

Mitochondrion

Smooth endoplasmic reticulum

Nucleus

Lysosome

Cytosol (the aqueous phase between organelles)

Peroxisome

A

The disruption is usually carried out in an isotonic medium (this medium may be a salt solution but is often 0.25 M sucrose).

After homogenization the components can be separated.

The fragments of the endoplasmic reticulum tend to fuse to form vesicles.

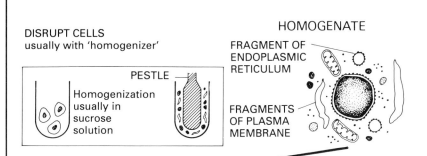

DISRUPT CELLS
usually with 'homogenizer'

PESTLE

Homogenization usually in sucrose solution

HOMOGENATE

FRAGMENT OF ENDOPLASMIC RETICULUM

FRAGMENTS OF PLASMA MEMBRANE

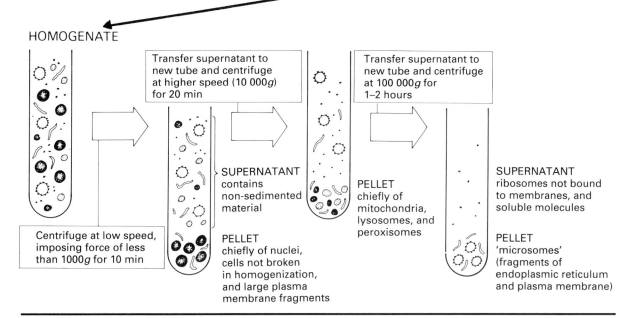

HOMOGENATE

Centrifuge at low speed, imposing force of less than 1000*g* for 10 min

Transfer supernatant to new tube and centrifuge at higher speed (10 000*g*) for 20 min

SUPERNATANT contains non-sedimented material

PELLET chiefly of nuclei, cells not broken in homogenization, and large plasma membrane fragments

Transfer supernatant to new tube and centrifuge at 100 000*g* for 1–2 hours

PELLET chiefly of mitochondria, lysosomes, and peroxisomes

SUPERNATANT ribosomes not bound to membranes, and soluble molecules

PELLET 'microsomes' (fragments of endoplasmic reticulum and plasma membrane)

B

Marker enzymes.

Confirmation of the nature of the morphological constituents of the various subcellular fractions isolated by differential centrifugation can be obtained either by electron microscopy of the pellets or by biochemical analysis. Thus, one can determine the DNA/protein or RNA/protein ratio. Enzymic analysis is also useful, based on the principle that one enzyme is only to be found associated with one particular morphological constituent of the cell (usually but not always true) and that the enzymic make up of a particular constituent is unique (e.g. all mitochondria are identical, probably true). Such enzymes are known as *marker enzymes*. Examples are glucose 6-phosphatase for endoplasmic reticulum and succinic dehydrogenase for mitochondria.

A

The morphological constituents of a typical microsome fraction of liver as revealed by the electron microscope.

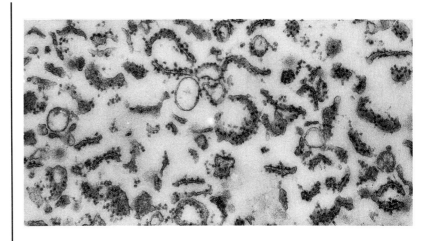

The microsome fraction is obtained by disrupting a cell and separating the components by differential centrifugation. That fraction which sediments more slowly than the mitochondrial fraction but is particulate is defined as the microsomal fraction. In liver extracts it consists predominantly of fragments of the endoplasmic reticulum. This is not necessarily the case with other types of cell and in all cases it is necessary to determine the morphological components of the microsome fraction before meaningful comparisons may be made.

B

Metabolic functions of organelles

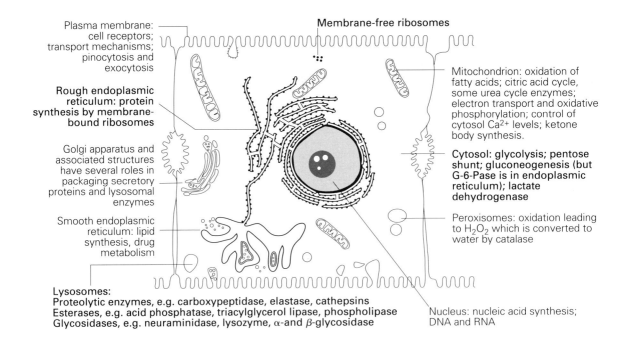

Plasma membrane: cell receptors; transport mechanisms; pinocytosis and exocytosis

Rough endoplasmic reticulum: protein synthesis by membrane-bound ribosomes

Golgi apparatus and associated structures have several roles in packaging secretory proteins and lysosomal enzymes

Smooth endoplasmic reticulum: lipid synthesis, drug metabolism

Membrane-free ribosomes

Mitochondrion: oxidation of fatty acids; citric acid cycle, some urea cycle enzymes; electron transport and oxidative phosphorylation; control of cytosol Ca^{2+} levels; ketone body synthesis.

Cytosol: glycolysis; pentose shunt; gluconeogenesis (but G-6-Pase is in endoplasmic reticulum); lactate dehydrogenase

Peroxisomes: oxidation leading to H_2O_2 which is converted to water by catalase

Lysosomes:
Proteolytic enzymes, e.g. carboxypeptidase, elastase, cathepsins
Esterases, e.g. acid phosphatase, triacylglycerol lipase, phospholipase
Glycosidases, e.g. neuraminidase, lysozyme, α-and β-glycosidase

Nucleus: nucleic acid synthesis; DNA and RNA

THE PROTEINS

The fundamental component of a protein is the polypeptide chain composed of amino acid residues. These may be covalently modified, other components being called prosthetic groups. The structure of the amino acids and their characteristic property as amphoteric molecules is described, followed by a summary of the way in which the amino acid residues interact within proteins. After showing how the order of the amino acids in polypeptides can be determined, the hierarchies of protein structure are illustrated. The correct tertiary structure is essential if the protein is to fulfil its biological role. This is shown by the effects of mutations. Proteins may be analysed by various means, the results being useful in the diagnosis of disease. Peptides containing as few as three amino acids may possess impressive biological activity; their structure and synthesis are described. The chapter ends with a description of the immunoglobulins and muscle proteins.

1. THE CHEMISTRY OF AMINO ACIDS AND PEPTIDES

Structure of amino acids

Amino acids have the general structure.

$$R$$
$$|$$
$$^+H_3NCHCOO^-$$

The internationally approved three letter and single letter abbreviations for each amino acid are indicated.

All the common amino acids except for proline have the same general structure, in that the α-carbon bears a –COOH and an –NH_2 group, but differ with respect to their 'R'-groups. The R-groups confer on the amino acids their respective characteristic properties. The amino acids are grouped according to the nature of their R groups as follows.

1. Non-polar or hydrophobic R-groups

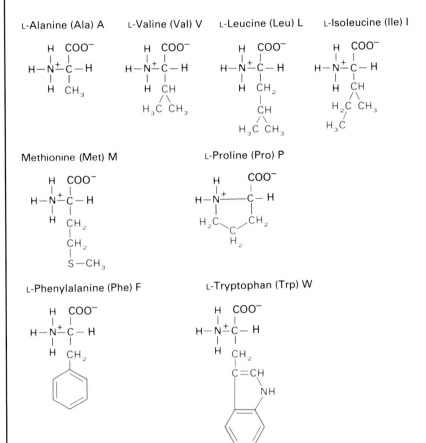

The α-carbon is optically active in α-amino acids other than glycine. The two possible isomers are termed D and L. All naturally occurring amino acids found in the proteins are of the L-configuration (see p 10).

2. Negatively charged R-groups at pH 6–7

3. Uncharged or hydrophilic R-groups

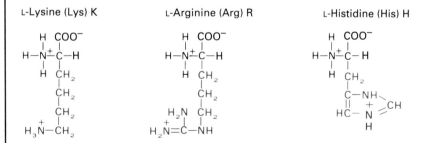

A cystine residue is formed from two cysteines linked through a disulphide bridge (-S-S-) formed from their sulphydryl (-SH) groups.

4. Positively charged R-groups at pH 6–7

The terms 'acid' and 'base' as applied to amino acids.

Acids are defined as proton donors and bases as proton acceptors. It follows that at pH 6–7, as shown, an amino acid in group 2 is present as a free base (an anion) and one in group 4 as a free acid (a cation). The terms 'acidic' and 'basic', as applied to amino acids, should therefore be used with caution, since they refer to the protonated forms of group 2 or the unprotonated forms of group 4. A compound such as an amino acid which carries both basic and acidic groups is referred to as *amphoteric*.

A

Asymmetry in biochemistry

Chirality from the Greek word *cheir* for 'hand'— the left and right hands are mirror images of each other.

Asymmetry in molecular structure is of great importance in biochemistry. A chiral molecule possesses at least one asymmetric centre, that is, a carbon atom to which are joined four groups different from each other.

The amino acid alanine may exist in two forms, denoted D-alanine and L-alanine.

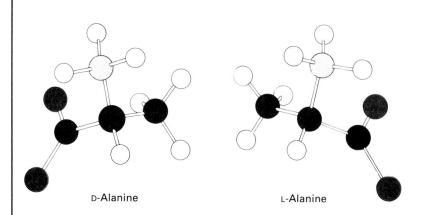

D-Alanine   L-Alanine

The amino acids contained in mammalian proteins are of the L-form. Sugars also are chiral molecules. D-Sugars predominate in mammalian carbohydrates (see p. 142). Red denotes carboxyl group oxygen atoms, grey is the amino group nitrogen, black is carbon and white is hydrogen.

B

Non-chiral asymmetry.

```
        COOH
         |
      H-C-OH     Plane of
    --+--+--+--  symmetry
      H-C-OH
         |
        COOH
```

meso-Tartaric acid

Even if a molecule is not chiral, it may contain identical groups which are nevertheless sterically distinguishable. The classic biochemical example is citric acid. Although this has a plane of symmetry, the centre carboxyl group and the hydroxyl group can be held in such a way that the two $-CH_2COOH$ groups can be distinguished (see p. 178). This can be seen in a simplified representation of a hypothetical molecule, which interacts with an enzyme with specific binding sites for different groups in the molecule.

If A and B are held in space on a surface, then the identical groups X_1 and X_2 can be distinguished. Such a molecule is termed 'pro-chiral' in that it can be made chiral by changing a group on only one of the central carbon bonds.

Note: If a molecule has a plane of symmetry such that chiral centres on either side of the plane of symmetry exactly compensate, then the molecule is termed a *meso* compound.

C

R and S convention.

A chiral centre can be denoted *R* or *S*.

The method for ascribing the *R*- or *S*-designation to a centre is as follows:

1   List functional groups in order of priority (see examples of the priority sequence in the list). Then orientate the molecule so that the group of the lowest priority points away from the observer.
2. If the order of priority (high to low) of the remaining groups is clockwise, the centre is *R*. If anticlockwise, it is *S*.

Thus, the α carbon of L-alanine (above) has the *S*-configuration.

The order of priority of some biochemically important groups is –SH (highest), –OH, –NH₂, –COOH, –CHO, –CH₃, –H (lowest).

A

Ionic properties of amino acids

The amphoteric★ properties of amino acids account for their separation on electrophoresis on paper at pH 6.0.

The amphoteric nature of α-amino acids determines that, in the absence of other acids or bases, the carboxyl and amino groups are both ionized fully, giving rise to the term 'zwitterion' (German *Zwitter* = hybrid, or hermaphrodite).

★ Property of behaviour either as an acid or a base.

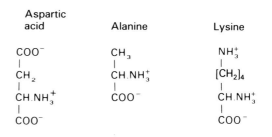

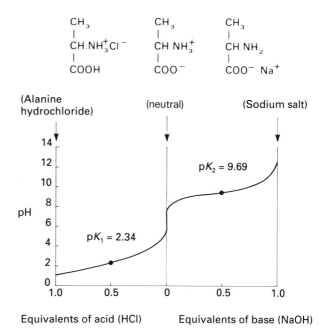

B

The buffering capacity of alanine is shown by titration with acid and alkali.

pK is defined on page 72.

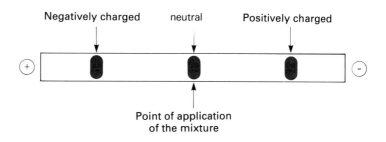

A

Histidine is an important amino acid in proteins since it contributes to their buffering capacity.

The basic groups of histidine may partially ionize at physiological pH.

The imidazole group of histidine is only weakly basic having a pK$_a$ of 6.00 and therefore exists as a mixture of the protonated and dissociated forms in solution at physiological pH.

The titration curve of histidine

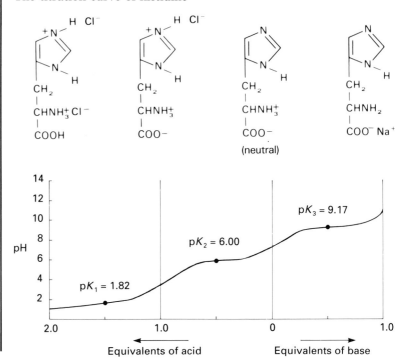

B

Amino acids can be separated by ion exchange chromatography.

In addition to permitting the separation of amino acids by electrophoresis, their ionic properties also permit their separation by ion exchange chromatography. Ion exchange is performed using a resin to which positively charged groups (anion exchange resin) or negatively charged groups (cation exchange resin) are covalently bound and thus immobilized. Ions passed down a column of such a resin bind competitively to the charged groups.

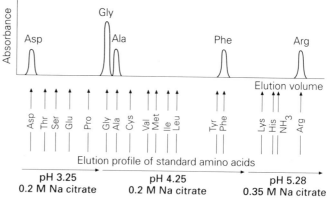

The figure shows the separation of the amino acids present in a peptide hydrolysate on a column of sulphonated polystyrene. The buffers of increasing pH cause aspartate (acidic) to emerge as the first amino acid and arginine (basic) as the last.

A

The peptide bond

The structure of peptides.

The peptide bond is formed by the interaction of two amino acids with the elimination of water between the neighbouring $-NH_2$ and $-COOH$ groups.

$$-CH \overset{R_1}{\underset{}{|}} - \underset{O}{\overset{||}{C}} - \underset{H}{\overset{|}{N}} - CH \overset{R_2}{\underset{}{|}} - \underset{O}{\overset{||}{C}} - \underset{H}{\overset{|}{N}} - CH \overset{R_3}{\underset{}{|}} -$$

Proline can also participate in a peptide bond

$$-HN - CH \overset{R}{\underset{}{|}} - \underset{O}{\overset{||}{C}} - N \underset{CH_2}{\overset{|}{}} - CH \underset{CH_2}{\overset{|}{}} - CO -$$
$$CH_2$$

but there is no H available on the $\underset{O}{\overset{||}{C}} - N -$ for hydrogen bonding.

Sharing of electrons between the $-\overset{|}{C}=O$ and $-C-NH-$ bonds confers rigidity on the peptide bond.

$$-C \overset{}{\underset{O}{\overset{\ldots}{|:}}} NH -$$

This has important implications for protein structure (see p. 20B).

B

The Biuret reaction.

Biuret has the formula $NH_2CONHCONH_2$ and, therefore, is a simple substance containing a peptide bond. When biuret is treated with $CuSO_4$ in alkaline solution a purple colour is produced. This is known as the *biuret reaction* and proteins give a strong reaction.

C

Notation used for peptides.

H₂N–Tyr–Gly–Gly–Phe–Met–COOH
Enkephalin

In writing the primary structure of a peptide one starts with the amino terminus* so that in the three-letter code enkephalin, for example, is abbreviated as above or in the single-letter code as YGGFM.

*The amino terminus is often abbreviated to N-terminus, and the carboxy terminus to C-terminus.

Non-covalent bonds in proteins

The R-groups interact in a *polypeptide chain* to create the tertiary structure of a protein.

Proteins are composed of chains of amino acids linked by peptide bonds. These are termed polypeptide chains.

The variety of bonds or interactions which stabilize the tertiary structure of protein molecules is shown. The S–S bonds formed by the oxidation of two sulphydryl groups are *covalent*. These are particularly likely to be present in proteins exposed to an unfriendly environment to enhance their rigidity. The other interactions are *non-covalent*. These may be either *apolar*, i.e. hydrophobic, or *polar*, i.e. hydrogen-bonding and ionic.

Hydrophobic interactions may be due to: (1) van der Waal's interactions which arise from an attraction between atoms due to fluctuating electric dipoles originating from the electronic cloud and positive nucleus; (2) the hydrophobic effect which refers to the tendency of non-polar groups to associate with one another rather than to be in contact with water.

Hydrogen bonds. These arise because when H is linked to O or N there is a shift of electrons that leads to a partial negative charge on the other atom. This produces an electric dipole that can interact with dipoles that exist elsewhere. The commonest hydrogen bond is between N–H and C=O as in the α-helix or β-pleated sheet (see pp. 20 and 21) but other bonds are possible as shown.

Ionic interactions or salt bridges are formed by the close approach of two atoms of opposite charge, e.g. between the glutamate and arginine residues as shown.

A

Ionic properties of peptides

A 'basic' protein with a predominance of group 4 amino acids is a cation at pH 7 and an 'acidic' protein with a predominance of group 2 amino acids is an anion at pH 7.

glycyl-aspartyl-lysyl-glutamyl-arginyl-histidyl-alanine

Illustrated is a hypothetical polypeptide containing all the R groups that normally contribute charges to proteins. The numbers represent the pK range of each dissociating group. Histidine is very important since its charge may vary over the physiological range.

B

Physical properties and separation of proteins

Proteins as electrolytes.

The isoionic point is the pH that results when the protein freed of all other ions is dissolved in water.

Isoelectric points of some common proteins

Blood proteins		Miscellaneous proteins	
Protein	Isoelectric point	Protein	Isoelectric point
α_1-Globulin	2.0	Pepsin	1.0
Haptoglobin	4.1	Ovalbumin	4.6
Serum albumin	4.7	Insulin	5.4
γ_1-Globulin	5.8	Histones	7.5–11.0
Fibrinogen	5.8	Ribonuclease	9.6
Haemoglobin	7.2	Cytochrome c	9.8
γ_2-Globulin	7.4	Lysozyme	11.1

The isoelectric point is the pH value at which there is zero migration to either electrode. According to the isoelectric point, proteins are described as basic, neutral or acidic depending on whether their overall charge at physiological pH is positive, approximately zero, or negative.

C

The titratable groups of a protein (ribonuclease).

Group	Amino acid involved	No. of residues	pK
α-NH$_2$	N-terminal	1	7–8
Side-chain COOH	Asp, Glu	10	4–5
Side-chain NH$_2$	Lys	10	10–11
Guanidyl	Arg	4	12–13
Imidazole	His	4	6–7
α-COOH	C-terminal	1	3–4

A

Methods for sequence determination in proteins

The primary structure of a protein can be determined either by the use of proteolytic enzymes or by the use of specific chemical reagents or both.

In a typical investigation, the N-terminus is determined by dansyl chloride, fluorodinitrobenzene or the Edman degradation technique. The Edman degradation can be used directly on the polypeptide chain to determine the amino acid sequence, as the amino acids are released one by one from the N-terminus. In other cases, the protein is split into peptides, either by CNBr to break the chain at methionine residues or by proteolytic enzymes acting at specific amino acid residues (see p. 211). The amino acid sequence of the resulting peptides is then determined, usually by the Edman degradation. The objective is to obtain peptides by the various methods that are overlapping in sequence, thus enabling the order in which individual peptide sequences occur in the protein to be determined. Since it is now possible to determine the base sequence of genes, the amino acid sequence may in certain cases be checked against the base sequence of the structural gene using the genetic code.

B

Chemical reagents.

Only one reagent is used commonly to break a polypeptide chain at a specific amino acid and this is *cyanogen bromide* ($N\equiv C-Br$), which cleaves adjacent to methionine residues.

Methionine

Homoserine lactone

C

Reactions at the N-terminus.

Another procedure is to react the N-terminal residue either with *fluorodinitrobenzene* or *dansyl chloride*. After reaction, the peptide is hydrolysed with acid and the dinitrophenyl amino acid or dansyl amino acid (fluorescent) identified.

Dinitrophenyl-labelled peptide (yellow)

Dansyl chloride
(5-Dimethylaminonaphthylsulphonyl chloride)

A

The *Edman degradation*.

In the Edman degradation the procedure is repeated on the ever-shortening chain and the procedure can be automated.

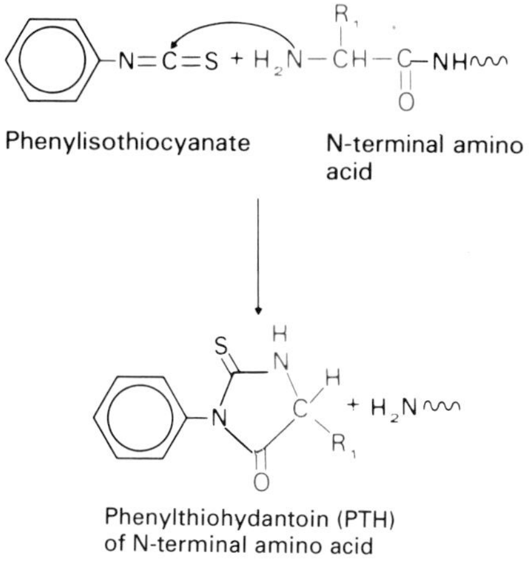

Phenylisothiocyanate

N-terminal amino acid

Phenylthiohydantoin (PTH) of N-terminal amino acid

The PTH of the amino acid can be identified.

Phenylisothiocyanate is used for the step-wise degradation of a peptide.

The labelled N-terminal residue (PTH-alanine in the first round) can be released without hydrolysing the rest of the peptide. Hence, the N-terminal residue of the shortened peptide (Gly-Asp-Phe-Arg-Gly) can be determined in the second round. Three more rounds of the Edman degradation reveal the complete sequence of the original peptide.

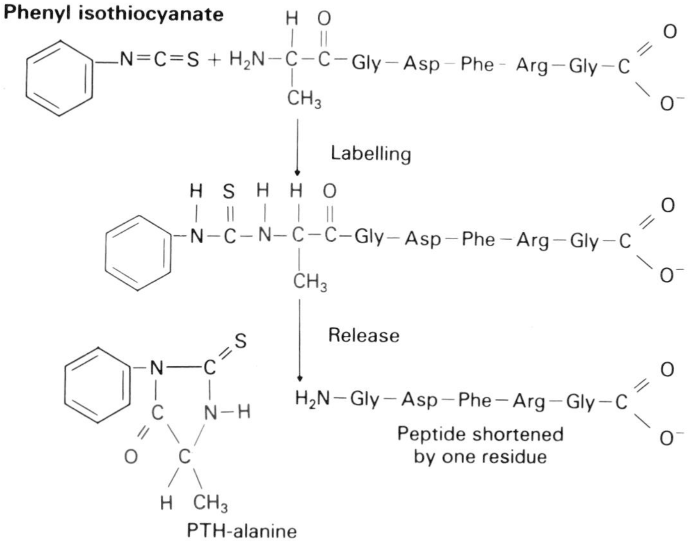

Phenyl isothiocyanate

Labelling

Release

Peptide shortened by one residue

PTH-alanine

B

The use of proteolytic enzymes.

The specificity of the various proteolytic enzymes that may be used to degrade a protein to peptides, the structure of which may then be determined, is shown on page 211.

As an example the action of trypsin is indicated below. Trypsin hydrolyses polypeptides on the carboxyl side of arginine and lysine residues.

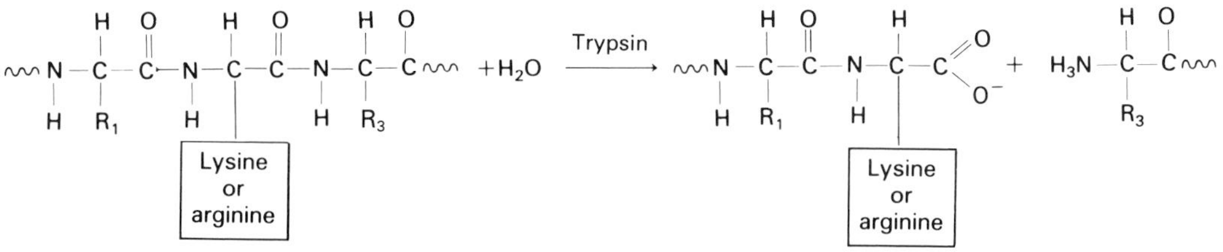

Lysine or arginine

Lysine or arginine

2. PROTEIN STRUCTURE, PROPERTIES AND SEPARATION

Primary structure describes the order of covalently linked amino acid residues.

Single amino acid residue

A

Hierarchies

The structure of a protein can usefully be considered in four hierarchies.

B

Role of hydrogen bonds.

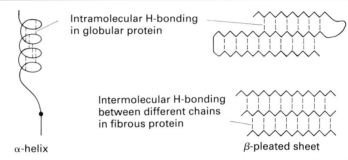

Intramolecular H-bonding in globular protein

Intermolecular H-bonding between different chains in fibrous protein

α-helix

β-pleated sheet

Secondary structure describes the way in which certain lengths of polypeptides interact through CO:NH hydrogen bonds either intramolecularly or intermolecularly.

C

Both the β-pleated sheet and α-helix play a role in tertiary structure.

β-pleated sheet

α-helix

Area of folding with no regular secondary structure (random coil)

Tertiary structure describes how the chains with secondary structure further interact through the R-groups of the amino acid residues to give a three-dimensional shape.

D

A protein with a quaternary structure is composed of several subunits. Such a protein is an *oligomeric* protein.

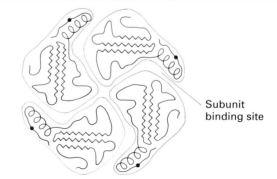

Subunit binding site

Quaternary structure describes the interaction, through weak bonds, of the polypeptide subunits. *Dimers* are associations of two subunits, *tetramers* are associations of four subunits, each subunit being a *monomer*. Multisubunit proteins with all subunits identical are known as *homo-dimers*, *homo-tetramers*, etc; those where they differ are *hetero-dimers*, *hetero-tetramers*, etc.

Protein denaturation and renaturation

For a protein to possess its unique biological activity it must possess the correct conformation.

By the application of heat and various reagents, a protein may lose its *native* form and become irreversibly *denatured*. An example is the denaturation of egg white when an egg is boiled. Denaturation is a common event during cooking; the resultant denatured protein being more susceptible to the action of proteolytic enzymes. *Reversible denaturation* may be achieved by the careful use of reagents such as urea and mercaptoethanol. Urea destroys the water structure and hence decreases the hydrophobic bonding of the R-groups of the amino acid residues (see p. 14), resulting in the unfolding and dissociation of the protein molecules. Mercaptoethanol reduces the S–S bonds. It may, therefore, be possible to renature the protein when the urea and mercaptoethanol are removed. The *renaturation* of ribonuclease is shown in the figure.

The reversible denaturation of ribonuclease is taken to indicate that a protein with the 'correct' primary structure will spontaneously fold to give the unique tertiary structure required for biological activity. This process is termed 'protein self-assembly'. It is now realized that there are two means whereby renaturation may be assisted. One involves *protein disulphide isomerase*, an enzyme that plays a role in 'correcting' wrongly paired S–S bonds. The other involves proteins named *molecular chaperones*. These are a family of unrelated classes of proteins that mediate the correct assembly of other polypeptides but are not components of the functional assembled structures. Examples are *heat shock proteins* and the *signal recognition particle* (see p. 113).

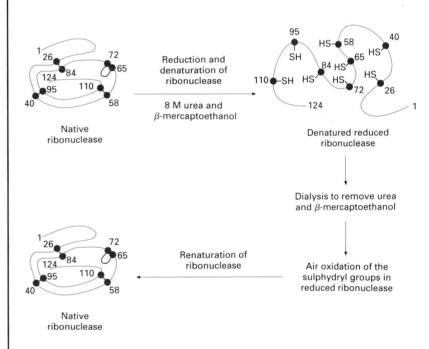

A

Nature of the peptide bond

Rotation in the peptide chain is limited.

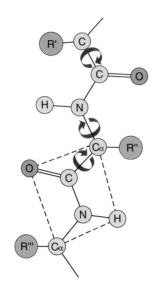

In a polypeptide chain rotation is only possible between certain atoms as indicated. The $C_\alpha CONHC_\alpha$ is planar.

B

The α-helix

Hydrogen bonds in the α-helix help to maintain its conformation.

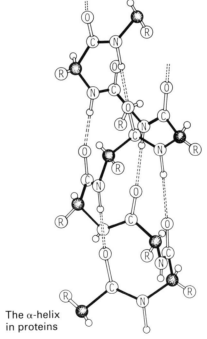

The α-helix in proteins

In the α-helix the maximum number of hydrogen bonds is present so that there are 3.6 amino acid residues per turn. In order for this to be achieved the polypeptide chain is twisted into a right-handed helix.

This structure is found not only in the fibrous proteins such as *keratin* but also in the tertiary structure of the globular proteins, as will be shown.

Because a proline residue in peptide linkage has no spare H to bond, proline has the effect of preventing α-helical formation.

A

The β-pleated sheet

Hydrogen bonds in the β-pleated sheet.

Within proteins the α-helices and β-sheets are connected by loop regions.

Polypeptide chains that are similar in conformation to those found in pleated sheets are said to have a β-conformation.

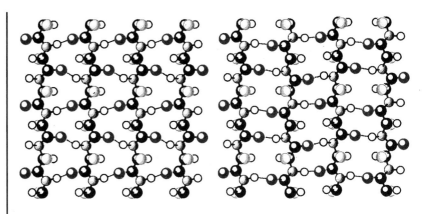

A polypeptide chain in which there are no intrachain H-bonds is described as in the β-conformation. Such chains may interact as shown. The R-groups are in the opposite plane to the hydrogen bonds and so do not interact. The chains may either be parallel, i.e. with the chains NH_2 to COOH running in the same direction (left) or antiparallel (right). The β-pleated sheet structure is not only found in fibrous proteins but is also present in the tertiary structure of globular proteins as will be illustrated.

B

Folding of globular proteins

The polypeptide chain of a globular protein folds so that apolar residues tend to be buried to form a hydrophobic interior with polar groups situated at the hydrophilic surface of the protein.

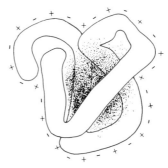

Calculations have been made to determine that fraction of particular residues for which more than 95% of the surface is buried. Twelve different proteins were studied and the average values for several amino acids is shown below:

Ileu 0.60, Val 0.54, Phe 0.50, Leu 0.45, Cys 0.40, Trp 0.27, Thr 0.23, Ser 0.22, Glu 0.18, His 0.17, Tyr 0.15, Asp 0.15, Lys 0.03, Arg 0.01

A characteristic of enzymes (see later) is that often they have a cleft extending into the protein interior which may contain some polar groupings.

A

The architecture of myoglobin (and of the subunits of haemoglobin, which have a similar structure) involves eight helical regions. These helical regions in both myoglobin and haemoglobin are denoted by the letters A to H, and the regions joining helices by pairs of letters (e.g. FG, joining helices F and G).

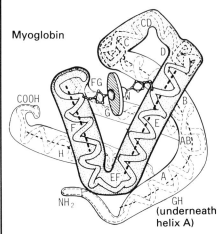

Histidines in helices E and F interact with haem on either side. The O_2 molecule sits at W. Helices E and F form the walls of a box for the haem, B, G and H are the floor and the CD corner closes the open end. The haem 'pocket' consists in the main of non-polar (group 1) amino acids.

Myoglobin carries charged amino acid residues at positions which are hydrophobic in haemoglobin and which are important in bonding the haemoglobin subunits together. Hence the myoglobin subunits do not readily associate to form a quaternary structure.

B

An example of the importance of quaternary structure can be seen in the way that the four chains of haemoglobin interact to give a compact structure, which is essential for correct protein function.

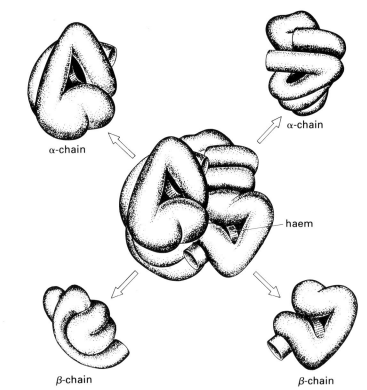

Human adult haemoglobin consists of two identical α-chains and two identical β-chains. If there is any change in the structure of the α and β-chains resulting from mutations that affect their interaction, the properties of the haemoglobin will be changed. Although this folding is common to the haemoglobins of every species possessing this type of molecule only about 10 amino acid residues are *invariant*.* Some changes in the primary structure make little difference to the tertiary structure whereas others have a profound effect, e.g. in *sickle-cell haemoglobin*.

*i.e. present in every case.

A

Myoglobin and haemoglobin structure

The four chains of haemoglobin are designated $\alpha_1\alpha_2\beta_1\beta_2$. There are few bonds between the two α-chains or between the two β-chains. However, there are strong hydrophobic bonds between unlike chains, e.g. $\alpha_1\beta_1$, $\alpha_1\beta_2$, $\alpha_2\beta_1$ or $\alpha_2\beta_2$. The bonds between an α- and a β-monomer are stronger than α–α or β–β bonds or those between one $\alpha\beta$ dimer and another. Thus if haemoglobin is dissociated, $\alpha\beta$ dimers always result.

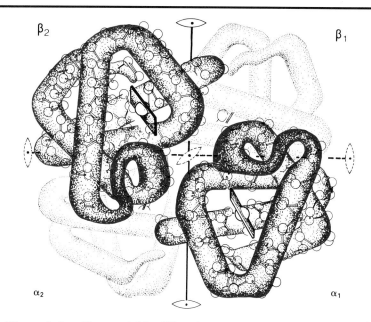

The α-chain of haemoglobin differs from the β-chain by the deletion of one residue in the NA segment, the addition of two residues in the AB corner and the deletion of six residues in the CD segment and D helix.

The binding of oxygen to haemoglobin results in the movements of α_1 and β_1 chains as a unit relative to the α_2 and β_2-chains. Deoxyhaemoglobin is denoted as having the T (tense) form, and oxyhaemoglobin the R (relaxed) form.

B

Treatment of haemoglobin with 8 M urea causes the molecule to dissociate into $\alpha\beta$ dimers rather than monomers or $\alpha\alpha$ or $\beta\beta$ dimers.

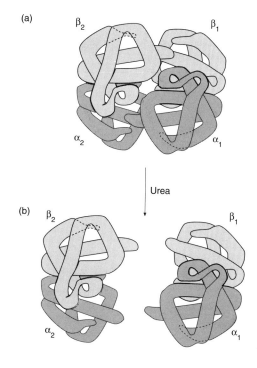

A

Oxygen binds to Fe^{2+} and this triggers the change in conformation.

Note that oxygen does not bind when the Fe^{2+} is oxidized to Fe^{3+} as is found in methaemoglobin.

The binding of oxygen to the iron atom reduces the diameter of the iron atom and it moves into the plane of the porphyin ring. The oxygen binds at a site near the distal histidine on helix E (see p. 22). The movement of the iron atom moves the proximal histidine F8 and produces a conformational change in that subunit. This is transmitted to other subunits.

Helix F

Proximal histidine F8

Fe

Oxygen or water

Distal histidine E7

B

Bisphosphoglycerate (BPG) binds specifically to deoxyhaemoglobin.

Bisphosphoglycerate (for its physiological role, see p. 26) binds to the central cavity created by the four subunits of deoxyhaemoglobin. It is extruded on oxygenation because this cavity becomes too small but tends to reduce the affinity for O_2.

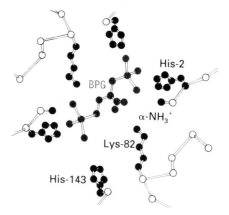

His-2

BPG

α-NH$_3^+$

Lys-82

His-143

The mode of binding of BPG to human deoxyhaemoglobin is by the interaction of BPG with three positively charged groups on each β-chain.

A

Myoglobin and haemoglobin properties

The difference in subunit structure of myoglobin and haemoglobin is reflected in their oxygen saturation curves.

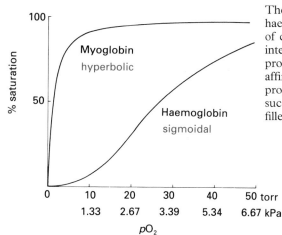

The sigmoidal curve of haemoglobin is indicative of cooperative interactions between protein subunits; the affinity for O_2 becomes progressively greater as successive O_2 sites are filled.

B

The effect of the pO_2 in capillaries and lungs is such that in the lungs haemoglobin becomes virtually fully saturated, whilst in the muscle capillaries it loses over 50% of its oxygen.

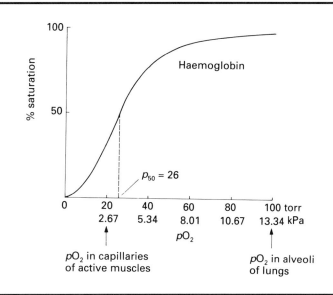

C

A lowering of the pH from 7.6 to 7.2 results in the release of O_2 from oxyhaemoglobin.

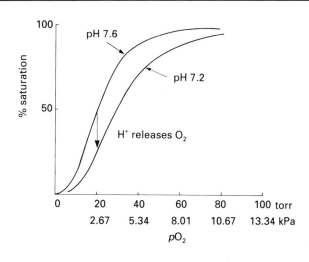

A

Bisphosphoglycerate (BPG) interacts with haemoglobin in the red blood cell.

Human red cells contain 2,3-bisphosphoglycerate (BPG). This substance reduces the affinity of haemoglobin for O_2. The amount of BPG varies with the physiological condition, being the same in mother and fetus. The affinity of BPG for fetal haemoglobin is much weaker than for the adult so that O_2 combines preferentially with fetal haemoglobin.

2,3–Bisphosphoglycerate

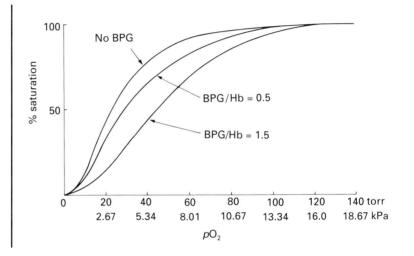

B

The presence of BPG reduces the affinity of O_2 for haemoglobin to a degree dependent on the BPG/Hb ratio.

To the extent that the binding of O_2 to haemoglobin is to be likened to the interaction of a substrate with an enzyme, BPG can be regarded as an allosteric effector (see p. 66).

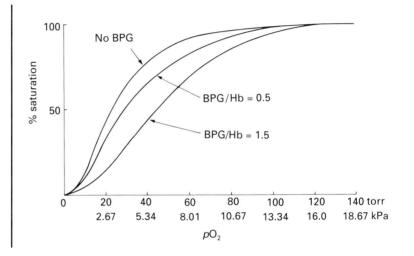

C

The effect of CO_2 is to decrease the amount of haemoglobin which is present in the oxygenated form. The effects of CO_2 and BPG are cumulative.

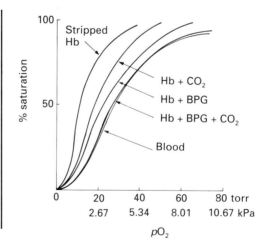

A

Abnormal haemoglobins

There are many cases in which one amino acid in one of the haemoglobin chains is replaced as a result of a single-point mutation in a globin gene.

Sickle-cell haemoglobin (HbS)

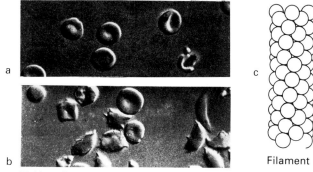

a
b
Sickle-cells

c

Filament

In HbS a glutamic acid in the β-chain (residue 6) is replaced by valine. This has little effect on the physiological properties of oxygenated HbS and the red blood cells appear normal (a), but deoxygenated HbS forms rod-like filaments (c) which cause the red cells to sickle (b).

B

Some replacements in different haemoglobins: residues participating in $\alpha_1\beta_1$ and $\alpha_1\beta_2$ contacts are shown in red (●); the positions of other variants which do not produce clinical symptoms are denoted half red (◐).

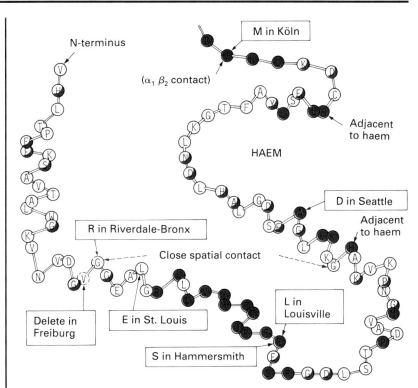

An invariant amino acid is one that has never been found to vary in any haemoglobin so far sequenced. There are only two amino acid residues that are common to all the globins: a histidine that forms a covalent link with the haem iron; and a phenylalanine in position 1 of the loop made by helices C and D which wedges the haem into its pocket. All other residues are replaceable, but in 33 specific positions replacements are restricted to non-polar residues.

A

Some abnormal haemoglobins have reduced oxygen affinity.

The effect on the saturation curve can be very marked.

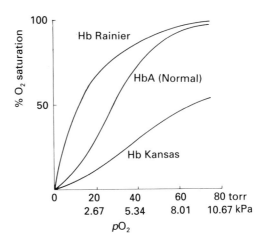

Substitution of an amino acid can cause increased or decreased oxygen affinity.

Some substitutions causing increased oxygen affinity

Haemoglobin	Substitutions	Site affected
Hb Rainier	β-145 (Tyr → Cys)	C-terminal
Hb Denmark Hill	α-95 (Pro → Ala)	$\alpha_1\beta_2$ contact
Hb Syracuse	β-143 (His → Pro)	BPG-β contact
Hb San Diego	β-109 (Val → Met)	Haem pocket

Many other abnormal haemoglobins have increased oxygen affinity.

Only a few haemoglobins are known which have reduced oxygen affinity. One of these is Hb Kansas, in which threonine is substituted for asparagine at β-102.

B

Unstable haemoglobins.

Note: FG5 β-98 indicates amino acid 98 in the β-chain, at position 5 in the FG region.

Other substitutions can cause instability of the haemoglobin molecule, leading to formation of methaemoglobin (HbM).

The cause of the instability of some unstable haemoglobins

Haemoglobin	Substitution	Cause of instability
Hb Köln	FG5 β-98 Val → Met	Large bulky side-chain of methionine distorts FG segment, breaking several haem contact amino acids
Hb Hammersmith	CD1 β-42 Phe → Ser	Phenylalanine is an important haem contact, the side-chain of serine being too short to reach haem group
Hb Bristol	E11 β-67 Val → Asp	Non-polar valine is replaced by polar aspartic acid; the internal siting of aspartic acid would result in gross distortion of the E helix.

A

Abnormal haemoglobins are detected either by a clinical abnormality or by screening by electrophoresis. In this regard, only those changes leading to altered electro-phoretic behaviour are detected. About 400 abnormal haemoglobins have been detected to date.

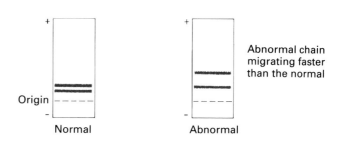

Although in the figure the abnormal haemoglobin moves further than normal, in other cases such as HbS it moves more slowly. This depends on the charge.

B

So-called glycosylated haemoglobin

Glucose may react with the N-terminal amino acid residues (valine in the human) of the β-chains of HbA to produce a so-called 'fast' haemoglobin A_1. The means of attachment is through a non-enzymic reaction so that a ketoamine is formed (glucose–CH_2–NH–βA). The reaction is a continuous process occurring slowly during the life span of the red blood cell over a period of about 120 days.

The concentration of HbA_1 is correlated to the degree of diabetic control so that a quantitative determination of HbA_1 reflects the patient's average blood glucose concentration over a long period.

HbA_1 is usually named glycosylated haemoglobin but this is a confusing nomenclature since the carbohydrate link is not glycosidic and is quite different from that of the glycoproteins. The name *glycated* haemoglobin is to be preferred.

The unglycated Hb, A_0, can be separated from glycated HbA_1 either by column chromatography or by electrophoresis as shown on agar gel. The charged groups of agar interact with A_0 to a greater degree than A_1, thus retarding the migration of A_0.

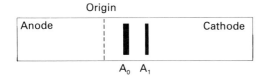

C

A note on the representation of molecular weights.

The term 'relative molecular mass' (symbol M_r) is preferred to 'molecular weight'. Both M_r and molecular weight are ratios and hence it is *incorrect* to give them units such as daltons (symbol Da). It is therefore incorrect to say that the M_r or the molecular weight of substance X is 10^5 Da. The dalton is a unit of mass equal to 1/12th the mass of an atom of carbon-12. Hence, it is *correct* to say the molecular mass of X is 10^5 Da or to use expressions such as the 16 000 Da peptide. For entities that do not have a definable molecular, weight it is *correct* to say, for example, 'the mass of a ribosome is 10^7 Da. A kilodalton (symbol kDa) is equal to 1000 Da.

A

Separation of proteins

Gel permeation chromatography

This technique utilizes a matrix, originally based on dextran that was cross-linked to form a mesh that only molecules of a certain size could penetrate (the greater the cross-linking, the smaller the holes of the mesh).

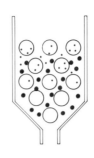
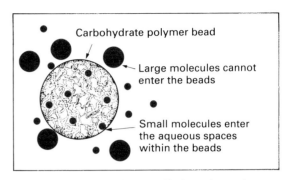

Large molecules which penetrate the mesh less readily have less volume through which to permeate, and thus elute faster. The matrix is normally packed in a column. The method can be used to separate small molecules, such as salts, from larger molecules, such as proteins, and to separate macromolecules (proteins, DNA) of different sizes. Nowadays, materials other than dextran, such as agarose, may be used as the basis of the matrix.

B

Sephadex★ chromatography of some serum proteins.

The red line indicates the separation of total protein in serum and the black line the separation of specific proteins (on different scales).

★A widely used dextran bead matrix.

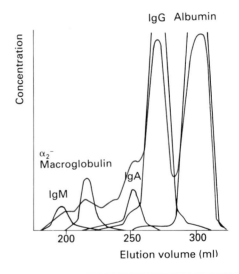

C

Gel permeation can be used for the determination of the M_r of a protein.

A plot of the elution volume V_e of native proteins of known M_r on Sephadex G-75 (●) and G-100 (○) versus log M_r is shown. (The grades of Sephadex differ in the extent of cross-linking of dextran.)

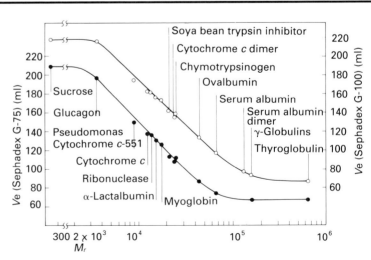

A

Electrophoresis

This may be performed on paper or cellulose acetate in buffer at pH 8.6. The proteins are stained after denaturation.

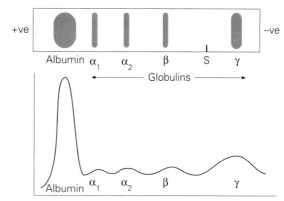

The figure shows the results of cellulose acetate electrophoresis of serum proteins from a normal human. The separated protein bands are visualized after staining with dye and a densitometric scan provides an indication of the relative amount of protein in each band. S shows the point of application of the serum before applying the current with the charges shown.

B

Polyacrylamide gel electrophoresis (PAGE).

PAGE can be used to separate native proteins according to their charge and size.

PAGE can also be carried out in the presence of sodium dodecyl sulphate (SDS). In this case, oligoproteins are separated into their subunits.

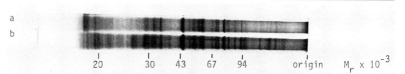

Above is shown the resolution of membrane proteins from (a) spleen lymphocytes and (b) thymocytes as a result of SDS–PAGE.

The proteins are suspended in a 1% solution of SDS. This detergent disrupts most protein–protein and protein–lipid interactions. Very often, 2-mercaptoethanol is added to disrupt disulphide bonds. The solution is layered on an acrylamide gel containing SDS and subjected to electrophoresis. The electrophoretic mobility of most proteins, but not glycoproteins, depends on their M_r rather than their net charge in the absence of SDS. The negative charge contributed by SDS molecules bound to the protein is much larger than the net charge of the protein itself. A pattern of bands appears when the gel is stained with Coomassie blue.

Agarose may be used in place of polyacrylamide for larger proteins or to obtain a different type of separation in the absence of SDS.

Two-dimensional PAGE can also be carried out by utilizing different conditions in each direction, e.g. an immobilized pH gradient (pH 4–7) in one direction and an 11–14% polyacrylamide gradient in the other.

A

Affinity chromatography

The specific ligand is covalently attached to the column matrix. Substances that bind to the ligand may then be passed down the column and are absorbed. They can be eluted by passing down the column solutions containing salts or other compounds that cause them to be desorbed.

Examples of types of ligand that have been used are shown.

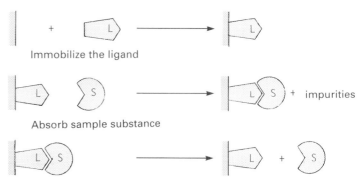

Immobilize the ligand

Absorb sample substance

Desorb bound substance

Enzyme:	Substrate analogue, inhibitor, cofactor
Antibody:	Antigen, virus, cell
Lectin:	Polysaccharide, glycoprotein, cell surface receptor, cell
Nucleic acid:	Complementary base sequence, histone, nucleic acid polymerase, binding protein
Hormone, vitamin:	Receptor, carrier protein
Cell:	Cell-surface-specific protein, lectin

B

High-performance liquid chromatography (HPLC)

In HPLC, the column packing is contained in a stainless steel column and the eluting buffer or solvent is pumped through the column under high pressure (up to 4000 p.s.i.). The packing is composed of small particles, sometimes as small as 3 μm in diameter, and very high resolution is obtained.

Peptide separation

1. Oxytocin
2. Met-Enkephalin
3. Thyrotropin-releasing hormone
4. α-Endorphin
5. Luteinizing-hormone-releasing hormone (LHRH)
6. Neurotensin
7. α-Melanocyte-stimulating hormone
8. Angiotensin II
9. Substance P
10. β-Endorphin

Conditions

Instrument:	Bio-Rad protein chromatography system
Column:	Bio-Gel TSK SP-5-PW
Sample:	2 μg each peptide in 50 μl
Eluant:	30 min linear gradient from A to B
	A. 0.02 M phosphate buffer of pH 3/CH_3CN (70/30)
	B. 0.5 M phosphate buffer of pH 3/CH_3CN (70/30)
Flow Rate:	1.0 ml min^{-1}
Temperature:	25°C
Detection:	UV at 220 nm

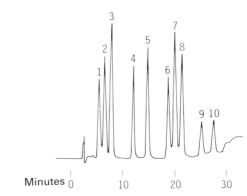

As an example of the high resolution of HPLC the separation of peptides on a cation exchange column is shown. The technique depends on the use of microfine column matrices to give high resolution rapidly. In addition to ion exchange, gel filtration and reversed-phase chromatography may be used.

A

Isoelectric focusing— separation, pI

Isoelectric focusing (IEF) is an electrophoretic technique that separates proteins according to their isoelectric point (p*I*) in a stable pH gradient generated by carrier ampholytes. These carrier ampholytes migrate under the influence of an electric current to generate the pH gradient, which increases from the anode to the cathode.

The figure shows the separation of proteins by horizontal, flat-bed isoelectric focusing using an ampholyte range from pH 3–10 (Pharmalyte 3–10): A, rainbow trout muscle extract; B, marker proteins, p*I* range 3–10; C, pike muscle extract. The sample is applied near the middle of the gel.

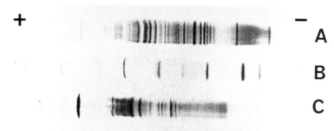

B

Large-scale separation of proteins

The scheme for plasma proteins shows how many different methods are used for the separation of some of the many plasma proteins. The products are analysed by gel electrophoresis.

Cryoprecipitation depends on the lesser solubility of some proteins in the cold.

Similar methods have been used in the isolation of serum components such as factor VIII for the treatment of haemophiliacs. Experience shows that a virus such as HIV may co-purify with the serum proteins. Heat treatment inactivates the virus.

A chromatographic procedure for plasma protein fractionation

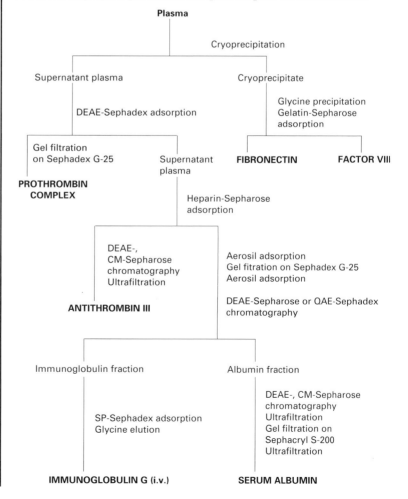

The pattern of serum proteins on electrophoresis may be used in the diagnosis of disease.

As explained previously, either paper, cellulose acetate or polyacrylamide gel may be used as the medium.

Globulin	Representative constituents
α_1	Thyroxine-binding globulin Transcortin Glycoprotein Lipoprotein Antitrypsin
α_2	Haptoglobin Glycoprotein Macroglobulin Ceruloplasmin
β	Transferrin Lipoprotein Glycoprotein
γ	γG γD γM γE γA

GLOBULINS

ALB α_1 α_2 β γ

Normal pattern

Primary immune deficiency

Impaired synthesis of immunoglobulins. Usually familial

Multiple myeloma

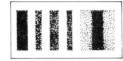

A monoclonal band, referred to as a paraprotein, which is due to production of a specific immunoglobulin by a malignant clone of cells. Paraproteins are also found rarely in other diseases

Nephrotic syndrome

Albumin lost into urine, and sometimes γ-globulin. Increase in α_2-globulin

Cirrhosis of liver

A polyclonal gammopathy with increase in many different immunoglobulins of all classes

Infection

Elevated α_1 and α_2 proteins. Usually decreased albumin. So-called 'acute-phase' response

Chronic lymphatic leukaemia

Quite often associated with decreased γ-globulin

Plasma should not be used:

If plasma is used instead of serum, fibrinogen band gives the appearance of a paraprotein, leading to misleading diagnosis

α_1 Antitrypsin deficiency

α_1-antitrypsin deficiency associated with emphysema of the lung in adults, and juvenile cirrhosis

3. PROTEIN FUNCTION

A

Serum proteins

Serum albumin

A transport protein (ligands in parentheses are transported by albumin when primary carriers are filled).

A ligand is a compound bound to a larger molecule.

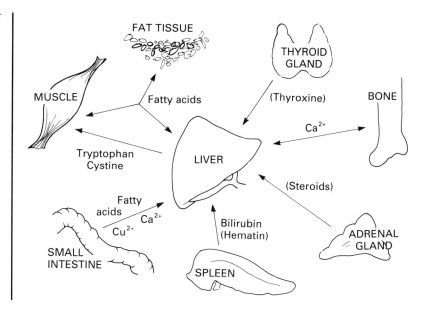

B

α_1-Antitrypsin and its role in emphysema

The predominant component of the α_1-globulin band consists of a protein named α_1-antitrypsin (AT). It is wrongly named because it is active against elastase rather than trypsin and is a member of a group of serine proteinase inhibitors or serpins (see pp. 63 and 64).

The dire effect of smoking

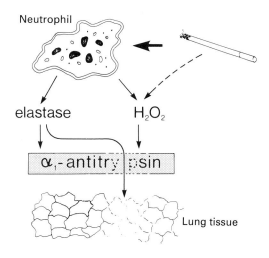

In normal lung the destructive power of elastase released from the neutrophils is held in check by AT. Cigarette smoking increases the number and activity of lung neutrophils and consequently the amount of elastase. Moreover, oxidation by H_2O_2 reduces the protection afforded by a given amount of circulating AT.

α_1-Antitrypsin is less active after oxidation and elastase then causes tissue breakdown and loss of elasticity in the lungs, i.e. emphysema.

A

The unhappy result of mutants of AT.

In addition to the normal protein there exist two variants, both confined to people of European descent: the S-variant in 5–8% (genotype MS) and the Z-variant in 4% (genotype MZ). The tissue specificity of AT depends on position 358 in the polypeptide chain of 394 amino acids being occupied by methionine (or valine). The methionine residue is susceptible to oxidation (see p. 35 B).

B

In non-smokers emphysema is unusual in people with either variant of AT.

Glu — Met

342   358

394 amino acids chain of normal AT

Elastase activity (+) properly controlled by circulating AT (−)

In the ZZ mutant, glutamine is replaced by lysine which results in only 15% of the AT being secreted. The retained AT accumulates in the endoplasmic reticulum of the hepatocyte where much is degraded, but some aggregates to form insoluble intracellular inclusions which are associated with the liver disease of ZZ children.

Lys — Met

Mutant Z AT

Elastase activity normal (+) reduction in circulating AT (−)

C

Smoking leads to tendency to emphysema.

Smoking not only increases neutrophil elastase but increases oxidation of Met.

Glu — Met oxid

Oxidation of AT involves Met–358

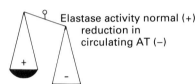

Elastase activity increased (+ circulating AT less active but amount normal (−)

D

Smoking in a mutant makes emphysema probable.

If the ZZ mutant smokes there is a reduction in the amount of circulating AT and it is less active.

Lys — Met oxid

Mutant Z AT in a smoker

Elastase activity is increased (+) circulating AT less in amount and less active (−)

Proteins involved in the metabolism of iron

Iron metabolism is regulated within tissues by the protein *ferritin* and transported in blood by the protein *transferrin*. Apoferritin is the iron-free form of ferritin. Ferritin catalyses the oxidation of Fe^{2+} to Fe^{3+}.

Transferrin is synthesized in the liver and present in the β-globulin fraction of serum. It is a glycoprotein with two iron (ferric) binding sites. Many cells possess transferrin receptors on which the cells depend for their essential supply of iron.

Ferritin, the iron store in liver and spleen, consists of 24 subunits and is a very large protein of M_r about 450 000. A thin coat of aggregated polypeptide monomers surrounds an inorganic core of up to 4500 iron atoms. Although ferritin is often thought of as a store for iron, its main function is as a means to detoxify iron. Hence its level in liver and spleen depends on the amount of iron to sequester.

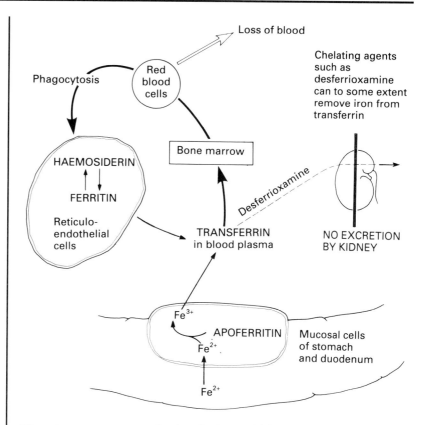

There is no excretory mechanism for iron, which slowly accumulates in the body throughout life, being absorbed in small quantity from the diet. Frequent loss of blood, such as in menstruation, may cause anaemia if dietary iron is inadequate, as iron is essential for haemoglobin synthesis, and thus red cell formation. When the red cells are destroyed, the iron is removed from haemoglobin by reticuloendothelial cells, where it remains until taken up by transferrin for utilization elsewhere. If excessive amounts of iron are absorbed, it may accumulate in the reticuloendothelial cells, such as the Kupffer cells of the liver, a condition known as iron overload or haemosiderosis due to the deposits of haemosiderin, a form of ferritin complexed with other proteins, and iron. The excretion of some iron through the kidneys may be induced by the chelating agent desferrioxamine, but loss of blood is the only really effective method of removing iron from the body. Ascorbic acid (vitamin C) strongly increases the absorption of iron.

Collagen

A connective tissue protein.

Collagen is the most abundant protein in mammals and is the major fibrous element of skin, bone, tendon, cartilage, blood vessels and teeth.

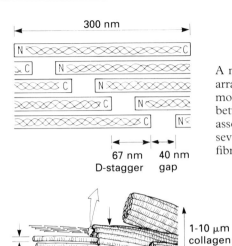

A microfibril is a staggered array of tropocollagen molecules. There is a gap between the ends. Microfibrils associate into a fibril and several fibrils form a collagen fibre.

In order to fulfil its many roles the basic structure of collagen is modified but always consists of three polypeptide chains of about 1000 amino acids, each in a left-handed helical conformation and wound around each other to form a right-handed supercoil. There are, in addition, covalent cross-links involving aldehydes derived from the lysine side-chains; such links may be either within or between the chains. Collagen from young animals lacks these cross-links and is called *tropocollagen*.

Depending on the amino acid composition of the three chains the collagens are classified into Types I to IV.

In contrast to globular proteins, the amino acid sequence of tropocollagen is regular except for about the last 20 amino acid residues at each end, and every third residue is glycine usually followed by proline. Two unusual amino acid residues are also present, hydroxyproline (Hyp) and hydroxylysine. Hence a typical sequence might be.

–Gly-Pro-Met-Gly-Pro-Arg-Gly-Leu-Hyp–

Each of the three strands is hydrogen bonded to the other two strands, there being no intrachain hydrogen bonds which are a feature of the α-helix. The H-donors are the peptide –NH groups of glycine and the acceptors are the peptide –CO groups of residues on other chains. Thus the direction of the H-bonds is transverse to the long axes of the tropocollagen rod.

The hydroxylation of the proline residues is effected by prolyl hydroxylase (for the role of ascorbic acid in this reaction, see p. 139).

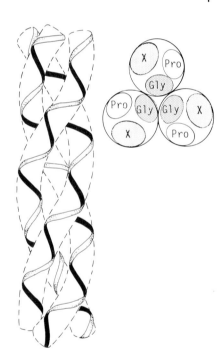

A model of the triple-stranded tropocollagen. To the right is a cross-section showing the outline of each strand and the position of glycine.

Peptides

Having described the importance of the tertiary structure of proteins, which is essential for their biological activity, it is a surprise to find that small peptides (sometimes as simple as a tripeptide) may have a very important and specific function, e.g. as a first messenger in neurotransmission and as a local mediator, or as a hormone. Such peptides vary in length from the three amino acids of thyrotropin-releasing hormone (TRH) to the 231 amino acids of human gonadotropin. Glutathione is an important tripeptide (see p. 165A).

The vasopressins are more correctly named *antidiuretic hormone (ADH)*. The particular N-terminal amino acid depends on the species of origin.

As indicated in the figure the smaller peptides often have a modified terminus. Thus in TRH the N-terminus is a cyclized glutamic acid (pyroglutamic acid) while there is an amide at the C-terminus. It is possible that such modifications enhance metabolic stability by protecting the peptides against exopeptidases. Another example of hormones produced in the posterior pituitary are oxytocin and vasopressin. The structures are shown in the figure and again the C-terminus is an amide.

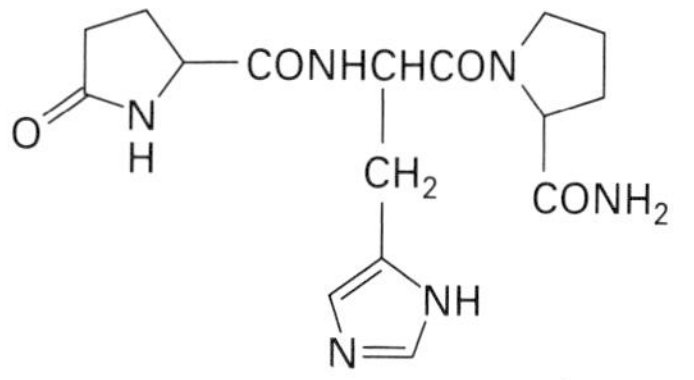

Thyrotropin-releasing hormone (TRH)
pyroglutamyl-histidinyl-proline amide

Cys-Tyr-Phe-Gln-Asn-Cys-Pro-Arg-Gly-NH$_2$
Arginine vasopressin

Cys-Tyr-Phe-Gln-Asn-Cys-Pro-Lys-Gly-NH$_2$
Lysine vasopressin

Cys-Tyr-Ile-Gln-Asn-Cys-Pro-Arg-Gly-NH$_2$
Oxytocin

Somatotropin (growth hormone) and prolactin are products of the anterior pituitary while placental lactogen is produced by the placenta during pregnancy. All three hormones are closely related in structure. Another group of hormones is the 'glycoprotein hormone family' which includes thyrotropin, follicle-stimulating hormone (FSH) and chorionic gonadotropin. They all contain numerous N-linked branched carbohydrate chains, and hence the name of the group.

The ACTH peptides

In the course of synthesis of many hormones the active peptide is split from a larger protein. An example of this is ACTH (adrenocorticotropic hormone).

A precursor protein that gives rise to more than one mature protein is known as a *polyprotein.*

β-Endorphin is an enkephalin. Although it can be produced as described, enkephalins are usually derived from preproenkephalin and preprodynorphin. The general structure of all the messenger RNAs (mRNAs) concerned shows strong similarities.

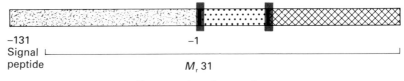

−131
Signal
peptide M_r 31

Prepro-opiomelanocortin

The initial translation product in the pituitary is a large protein containing a *signal peptide* at the N-terminus. (The significance of the signal peptide is explained on page 113.) The signal peptide is removed to give pro-opiomelanocortin with an M_r of 31 000. (The M_r values given in the illustrations represent thousands.) Cleavage of pro-opiomelanocortin takes place at sites which are characterized by the presence of two basic amino acid residues indicated by the red bands. A family of proteases (Kex2, PC2, PC3) are responsible for the cleavages. A similar process takes place in all specialized neuroendocrine cells where the hormone is stored within dense-core particles for later release. The processes involved in proinsulin conversion are indicated on p. 82).

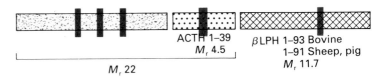

ACTH 1–39   βLPH 1–93 Bovine
M_r 4.5   1–91 Sheep, pig
                          M_r 11.7

M_r 22

There is further cleavage to produce various peptides. In particular, ACTH can be cleaved to α-melanocyte-stimulating hormone (αMSH). β-Lipotropin (βLPH) can be cleaved to β-endorphin (βEnd). The latter has morphine-like properties, in that it binds to opiate receptors in nervous tissue. γ-Lipotropin (γLPH) and 'corticotropin-like intermediate peptide' (CLIP) are also formed at this stage. βLPH was once thought erroneously to mobilize fatty acids from adipose tissue.

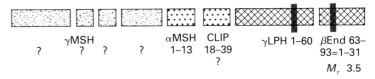

        γMSH   αMSH CLIP   γLPH 1–60 βEnd 63–
?   ?   ?   ?   1–13 18–39   93=1–31
                                          ? M_r 3.5

Subsequent protease activity yields 'β-melanocyte-stimulating hormone' (βMSH) and enkephalin (Enk). The latter also binds to opiate receptors.

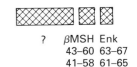

?   βMSH Enk
      43–60 63–67
      41–58 61–65

The numbers refer to the order of the amino acid residues starting at 1 for the N-terminus of ACTH or βLPH. Amino acid residues to the left of ACTH are indicated by a minus sign.

A

Immunoglobulin structures

The structure and function of immunoglobulin.

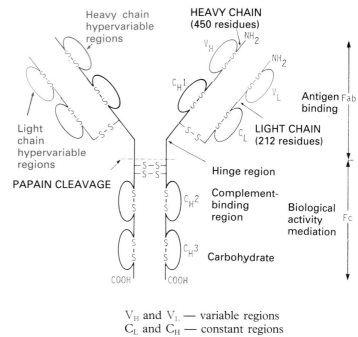

V_H and V_L — variable regions
C_L and C_H — constant regions

B

Three-dimensional structure of a human immunoglobulin.

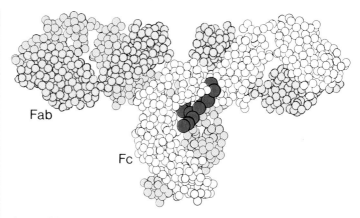

Space-filling representation of the three-dimensional structure of human IgG Dob. (Dob is human IgG1 (κ) cryoglobulin.) Dob has a 15-residue deletion in its hinge region as compared to normal IgG proteins. The two heavy chains are shown in *white* and in *pink*. Light chains are in *grey*. Carbohydrate is shown as *red* spheres.

C

The flexibility of the structure of IgG.

The shape of an IgG molecule can vary considerably so that the relative arrangement of the domains may change. The figure summarizes sites of movement. There are two peptides involved in these changes—the hinge peptide at the junction of C_H1 and C_H2 and the switch peptide at the junction of the V and C domains in Fab.

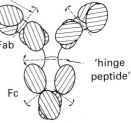

A

The various classes of immunoglobulins
Physical, chemical and biological properties of human immunoglobulin classes

Property	IgG	IgA	IgM	IgD	IgE
Usual molecular form	Monomer	Monomer, dimer, etc.	Pentamer	Monomer	Monomer
Molecular formula	$\kappa_2\gamma_2$ or $\lambda_2\gamma_2$	$(\kappa_2\alpha_2)_n$ or $(\lambda_2\alpha_2)_n$	$(\kappa_2\mu_2)_5$ or $(\lambda_2\mu_2)_5$	$\kappa_2\delta_2$ or $\lambda_2\delta_2$	$\kappa_2\epsilon_2$ or $\lambda_2\epsilon_2$
Other chains		J-chain, S-piece	J-chain		
Subclasses	IgG1, IgG2, IgG3, IgG4	IgA1, IgA2	None established	None	None
Subclass heavy chains	$\gamma1, \gamma2, \gamma3, \gamma4$	$\alpha1, \alpha2$	μ	δ	ϵ
M_r	150 000	160 000 and dimer	900 000	185 000	200 000
Sedimentation constant $(S_{20,w})$	6.65	7S, 9S, 11S	19S	7S	8S
Carbohydrate content (%)	3	8	12	13	12
Serum level (adult average)	8–16 mg ml^{-1}	1.4–4 mg ml^{-1}	0.5–2 mg ml^{-1}	0–0.4 mg ml^{-1}	17–450 ng ml^{-1}
Percentage of total serum immunoglobulin	80	13	6	0–1	2
Paraproteinaemia	Myeloma	Myeloma	Macroglobulin-aemia	Myeloma	Myeloma
Antibody valence	2	2, 4	5 (10)	2	2
Biological properties	Secondary Ab response; placental transfer	Characteristic Ab in mucous secretions	Primary Ab response	Lymphocyte cell surface molecule	Protection of external body surfaces

B

Antibody variants:

Isotypes
Allotypes
Idiotypes.

Antibody variants:

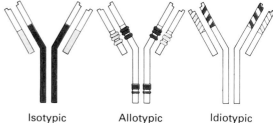

Isotypic   Allotypic   Idiotypic

Isotypic variation refers to the different heavy and light chain classes and subclasses (see above): the variants produced are present in *all* healthy members of a species. Allotypic variation occurs mostly in the constant region: *not all* variants are present in all healthy individuals. Idiotypic variation occurs in the variable region only and idiotypes are specific to each antibody molecule.

A

The formation of IgM polymer depends on the presence of the J-chain.

In IgM the hinge region in IgG is replaced by a rigid pair of extra domains (C_H2) while the C_H3 and C_H4 domains in IgM are structurally equivalent to the C_H2 and C_H3 regions respectively in IgG. IgM also possesses the J-chain, which is a polypeptide of M_r 15 000 rich in cysteine. The J-chain links two 7S monomers and this allows for the formation of S–S bridges between the C_H3 and C_H4 domains of other monomers, so forming a pentamer. Similarly in IgA, the J-chain is present. Each molecule of IgG has a valency of 2, i.e. has the potential to combine with two molecules of antigen (in theory IgM has a valency of 10 but the effective valency is about 5).

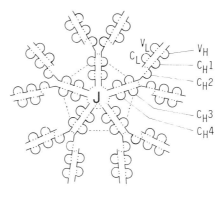

B

The valency of an antibody and an antigen may be defined as the number of sites of interaction. In most cases a given antigen will possess more than one antigenic site, as shown in the figure.

Below is shown the interaction of a *hapten*, bis-*N*-dinitrophenyl octamethylenediamine, with IgG. The two dinitrophenyl (DNP) groups are far enough apart not to interfere with each other's combination with antibody.

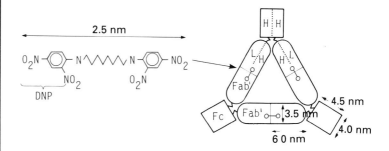

It is, of course, the interaction of antibody and antigen in this way that leads to precipitation.

Polyclonal antibodies

Most foreign substances when administered to an animal give rise to a mixture of antibodies each specific for a particular idiotype, i.e. a unique amino acid sequence that forms an antibody-combining site. A substance that gives rise to a polyclonal antibody in this manner is called an *immunogen*. Polyclonal antibodies, therefore, markedly differ from *monoclonal antibodies* (see p. 49).

The structure of complement

The structure of C1q is an example of a protein which combines the features of a globular and a fibrous protein.

Complement is the name given to a complex series of some 20 proteins which, along with blood clotting, fibrinolysis and kinin formation, forms one of the triggered enzyme systems found in plasma. These systems characteristically produce a rapid, highly amplified response to a trigger stimulus mediated by a *cascade* phenomenon where the product of one reaction is the enzyme catalyst of the next.

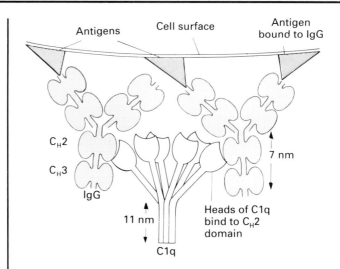

The Fc region of IgG is responsible for triggering pathways of the immune response that lead to the lysis of unwanted organisms. The *complement* system consists of a series of proteases. The first stage is triggered by the binding of a molecule, C1q (M_r 400 000), to the C_H2 domain of an IgG–antigen complex as shown.

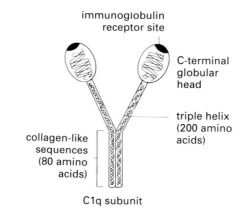

C1q structurally resembles a bunch of six tulips as shown. There are three chains in each tulip. The 'flower' is a globular structure while the 'stalk' is elongated and resembles the triple-helical structure of collagen (see p. 38). Indeed, it even contains the Gly-X-Y sequence with hydroxyproline.

A

The generation of antibody diversity

An animal can synthesize many millions of antibodies with different specificities. There are three sources of such diversity:

(1) A large repertoire of variable-region genes
(2) Somatic recombination, e.g. any of several hundred V genes can become linked to any of five J genes
(3) Somatic mutation, e.g. in the formation of λ light chains.

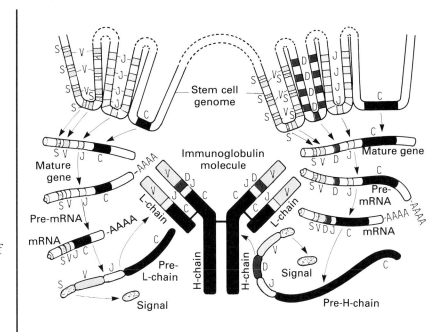

The stem cell genome contains multiple variants of the L-chain V and J genes and of the H-chain V, J and D (diversity) genes. The V genes are preceded by a small S segment coding for the signal peptide (see p. 112). As lymphocytes mature, each differentiating cell constructs particular L and H genes of virtually unique structure by a recombination process that randomly selects one out of each set of gene segments and assembles them together with a C gene. The pre-mRNA is spliced (see p. 121) to give a mature mRNA which is translated and the signal peptide removed. There is then oxidative formation of S–S bridges.

B

A glossary of some immunological terms useful to biochemists

Acquired immune deficiency syndrome
AIDS is defined by the Centers for Disease Control in the USA as 'a reliably diagnosed disease that is at least moderately indicative of an underlying immune deficiency or any other cause of reduced resistance reported to be associated with that disease'. AIDS is caused by the human T-cell lymphotrophic virus HIV-1

Adjuvant and Freund's adjuvant
A substance which non-specifically enhances the immune response to an antigen. A Freund's adjuvant is an emulsion of aqueous immunogen in oil: complete Freund's adjuvant also contains killed *Mycobacterium tuberculosis*, while incomplete Freund's does not

Bence Jones protein
Free immunoglobulin light chain dimers found in the serum and urine of patients and animals with multiple myeloma

β_2-microglobulin
A polypeptide which constitutes part of some membrane proteins including the class 1 MHC molecules

Cryoglobulin
An antibody or immune complex which forms a precipitate at 4°C

Cyclosporine
An immunosuppressive drug which is particularly useful in suppression of graft rejection

Fragments
Produced by chemical treatment:
H, L Heavy and light chains, which separate under reducing conditions
Fab Antigen-binding fragment (papain digestion)
Fc Crystallizable (because relatively homogeneous) fragment (papain)
F(ab')$_2$ Two Fab fragments united by disulphide bonds (pepsin)
pFc' A dimer of C_H3 domains (pepsin)
Facb An Ig molecule lacking C_H3 domains (plasmin)

Myeloma
A lymphoma produced from cells of the B-cell lineage

Tolerance
A state of specific immunological unresponsiveness

A

Immunoglobulin synthesis

IgM that is located on the plasma membrane differs in structure from that which is secreted.

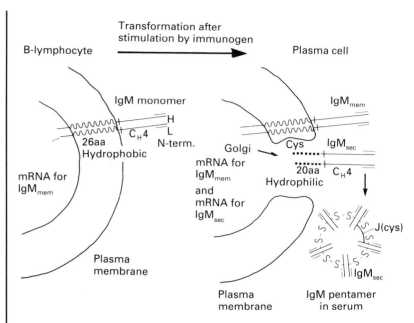

The figure compares the structure of mouse IgM that is retained by the plasma membrane (left) and that which is secreted to form a pentamer (right). IgM$_{mem}$ (membrane), IgM$_{sec}$ (secreted).

The B-lymphocyte bears a monomer of IgM (see p. 43 A) on its surface. When such a lymphocyte is recognized by an immunogen it is stimulated to form a clone of plasma cells which synthesize and secrete the IgM pentamer. It is now known that the structure of the C-terminal region of the H-chain accounts for the fact that the monomer IgM is an integral membrane protein (see p. 257) while the IgM pentamer is secreted. The monomer H-chain possesses a hydrophobic peptide of 26 amino acids which has an affinity for the plasma membrane while the H-chains of the pentamer have instead a hydrophilic peptide of 20 amino acids.

B

The immunoglobulin superfamily at the cell surface

MHC antigens are the major histocompatibility antigens recognized in graft rejection and function as determinants that are recognized along with foreign antigen in T-lymphocyte responses. The V and C inside ○ shows whether the domains are more like IgV or C regions. N-linked carbohydrate is shown as ⟶.

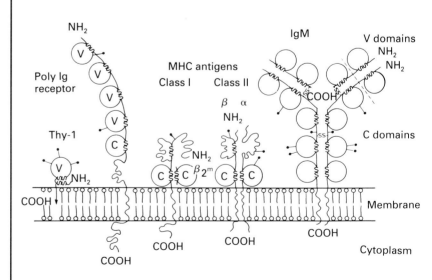

$\beta_2 m = \beta_2$–microglobulin

A

Immunoassay

The use of antibodies for the assay of substances of clinical significance.

Principles of radioimmunoassay

1. A limiting and constant amount of antibody 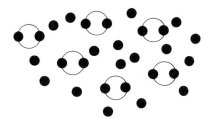 is incubated with a constant concentration of radioactively labelled analyte such that, for example, 40–50% of the labelled analyte is bound by the antibody.

2. On the addition of unlabelled analyte ●, competition occurs between labelled and unlabelled analyte for the limiting concentration of binding sites in the antibody.

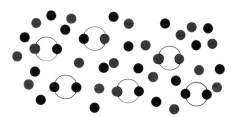

3. The amount of labelled analyte remaining bound to the antibody can be measured after separation of the antibody bound and unbound fractions. When no competing unlabelled analyte is present, the amount of labelled analyte bound will be high. When competing analyte is present, less labelled analyte will be bound to antibody.

B

Enzyme-linked immunoassay (ELISA)

In order to overcome the disadvantages of radioactivity, enzymes are used as markers.

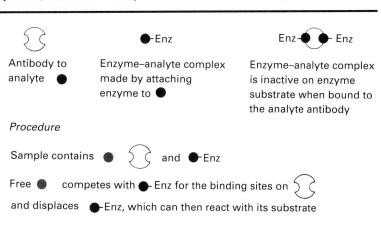

Quantitative estimation of the enzyme reaction shows how much enzyme was displaced, which is related to the amount of analyte.
Enzyme activity is measured by adding substrate for the enzyme. Often a dehydrogenase such as glucose 6-phosphate dehydrogenase, which requires NAD^+, is used so that the reaction can be assayed by spectrophotometry. Another useful enzyme is xanthine peroxidase.

*Enzyme-linked immunoassay
(ELISA) — cont.*

An alternative strategy as
applied in the Cambridge
Life Sciences plc assay of
progesterone in cow's milk.

1.

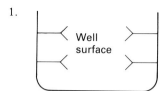

Progesterone antibody-binding
sites

Surfaces of wells of plate (solid phase) are precoated with progesterone
antibody, which binds through the Fc domain.

2.

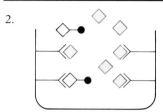

◇ Progesterone in milk sample or
standard

◇—● Enzyme-labelled progesterone

Enzyme-labelled progesterone competes for limited sites on antibody. The
enzyme is alkaline phosphatase.

3.

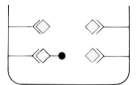

Unbound reactants removed by washing.

4.

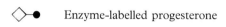
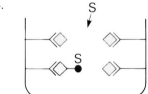

S, substrate to enzyme
(*p*-nitrophenyl phosphate)

Incubation followed by direct measurement of the optical density of the
solution or colour intensity. The amount of enzyme bound is inversely
proportional to the concentration of progesterone in the original sample.

Oestrus detection and pregnancy diagnosis are interpreted directly from a
standard curve.

Production of monoclonal antibodies

In the example, mice are immunized with an immunogen bearing two epitopes, *a* and *b*. The spleen cells make anti-*a* and anti-*b* which appear as antibodies in the serum. The spleen is removed and the individual cells fused in polyethylene glycol with constantly dividing (i.e. 'immortal') B-tumour cells selected for a purine enzyme deficiency.★ The resulting cells are distributed into micro-well plates in HAT (hypoxanthine, aminopterin, thymidine) medium which kills off all but the fused cells. Subsequent dilution of the medium ensures that each well contains one 'hybridoma' cell or less. Each hybridoma being the fusion product of a single antibody-forming cell and a tumour cell will have the ability of the former to secrete a single species of antibody, and the immortality of the latter, enabling it to proliferate continuously, clonal progeny providing an unending supply of antibody with a single specificity — the monoclonal antibody.

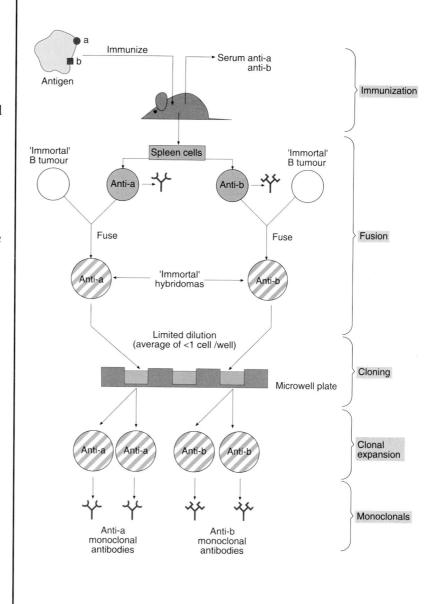

★The mutant myeloma cell line lacks hypoxanthine-guanosine phosphoribosyltransferase (HGPRT). This enzyme catalyses the synthesis of inosinate (a precursor of AMP and GMP) in the salvage pathway (see p. 96). In the HAT medium the aminopterin blocks de novo synthesis of nucleotides. Hypoxanthine cannot be used by the unfused myeloma cells because they lack HGPRT. Spleen cells contain HGPRT but are not 'immortal'. The hybridoma cells survive because they are both immortal and contain the HGPRT of the spleen cells.

A

Proteins of muscle

Striated muscle consists of fibres, which in turn are composed of fibrils. It is at the level of fibrils that the molecular level is approached.

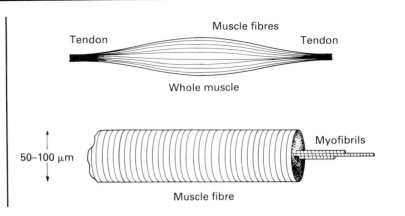

B

Within the muscle fibre, the fibrils are closely associated with the sarcoplasmic reticulum.

The A and I bands are responsible for the striated appearance under the microscope.

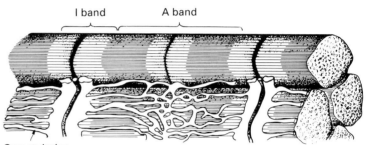

The sarcoplasmic reticulum contains one major protein, a Ca^{2+}-stimulated ATPase. When the muscle is stimulated, the sarcoplasmic reticulum releases Ca^{2+}, which is needed for the contractile process. Ca^{2+} is then pumped back into the sarcoplasmic reticulum by the Ca^{2+}-stimulated ATPase.

C

The myofibrils contain the protein myosin, and an assembly of the proteins actin, troponin and tropomyosin.

Actin is present not only in muscle cells but in all eukaryotic cells.

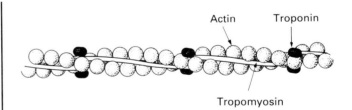

These assemblies of actin, troponin and tropomyosin are termed the *thin filaments*. These slide against the *thick filaments* of myosin during contraction.

D

The thick filaments are bundles of myosin molecules, with the myosin headpieces protruding from the side of the bundle.

A

The three-dimensional structure of myosin conduces to the formation of complexes from which the myosin headpieces protrude.

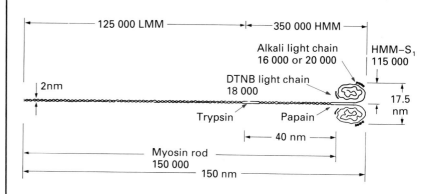

Red arrows indicate the points at which the myosin peptide chain can be cleaved by different agents. HMM, heavy meromyosin; HMM-S_1, heavy meromyosin headpiece; LMM, light meromyosin; DTNB, dithionitrobenzene.

B

Under the action of ATP and Ca^{2+}, the myosin headpieces interact with the actin–troponin–tropomyosin complex to initiate the contraction process.

In smooth muscle, which lacks striations, but contains both myosin and actin filaments, contraction is not regulated by the troponin-tropomyosin mechanism but by the degree of phosphorylation of its myosin light chains.

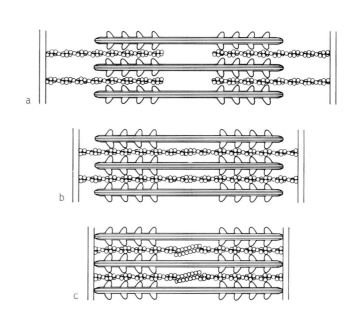

During contraction, the thin filaments (a) slide against the myosin headpiece until (b) they meet and (c) overlap. The headpieces of myosin contain ATPase activity that is associated with the mechanism by which the thick and thin filaments slide against each other.

A

The four classes of protein structures

X-ray crystallography of more than 580 different proteins has revealed common patterns.

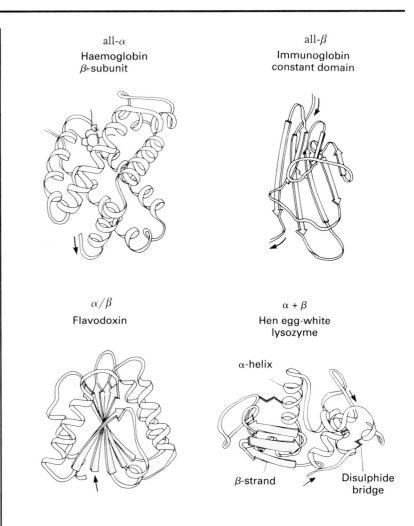

all-α
Haemoglobin
β-subunit

all-β
Immunoglobin
constant domain

α/β
Flavodoxin

$\alpha + \beta$
Hen egg-white
lysozyme

α-helix

β-strand

Disulphide
bridge

The classification is based on the relative order of α-helices and β-strands along the polypeptide chain. The α-helices and pleated (β) sheets are depicted as on page 65. In α/α structures there are mainly α-helices and little or no β-sheets. In β/β structures there are several β-strands but little or no α-helix. In IgG there are two β-sheets that pack together but they are twisted. In α/β structures α-helices and β-strands tend to alternate. Often the β-strands form a parallel sheet which is surrounded by α-helices. In $\alpha + \beta$ structures there are α-helices and β-strands that tend to segregate into different regions of the polypeptide chain. In the haemoglobin illustrated the haem is shown in red.

Summary of methods of investigation of the structure and function of proteins

Modern techniques can be synchronized to unravel the relationship between structure and function of both soluble and membrane proteins (see, for example, the section on membrane receptors, p. 259. The combination of techniques across the whole spectrum of biochemical knowledge (from molecular biology to physics) is one of the most significant advances in experimental capability in biochemical investigation.

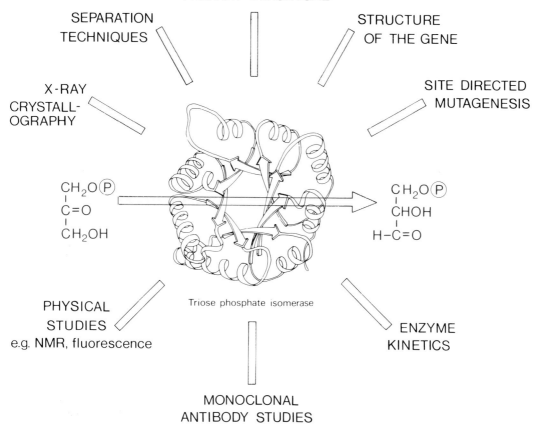

PRIMARY STRUCTURE

SEPARATION TECHNIQUES

STRUCTURE OF THE GENE

X-RAY CRYSTALL-OGRAPHY

SITE DIRECTED MUTAGENESIS

$$CH_2O\;\textcircled{P}$$
$$|$$
$$C=O$$
$$|$$
$$CH_2OH$$

$$CH_2O\;\textcircled{P}$$
$$|$$
$$CHOH$$
$$|$$
$$H-C=O$$

PHYSICAL STUDIES
e.g. NMR, fluorescence

Triose phosphate isomerase

ENZYME KINETICS

MONOCLONAL ANTIBODY STUDIES

Separation techniques. Protein purification remains an important approach to the investigation of protein structure. The aim is to purify the protein to homogeneity (absence of all contaminants). In the process, information is obtained regarding its size (gel permeation), charge (ion exchange) or isoionic point (isoelectric focusing). Enzymes such as trypsin can then be used to obtain peptides which can be sequenced, and from these sequences one can design oligonucleotides for use in screening complementary DNA (cDNA) libraries in the hope of selecting full-length clones of the gene coding for the protein.

Cloning techniques can further be used to determine the base sequence of the gene, and from this the amino acid sequence of the protein can be predicted. Site-directed mutagenesis can be used to prepare mutant genes which can be expressed to yield modified proteins.

Physical studies fluorescence, nuclear magnetic resonance (NMR) give data related to the dynamic properties of the protein in solution.

Enzyme kinetics can be used to analyse structure–function relationships with regard to the active site and allosteric sites.

Monoclonal antibodies can be prepared which aid detection of the protein by immunofluorescence, or permit preparation of affinity columns. The site to which a monoclonal binds can often be determined, leading to clues regarding the function of the binding site.

X-ray crystallography remains the only secure method for determining the three-dimensional structure of a protein, which then provides a basis for interpreting the other structural data.

3

THE STRUCTURE AND FUNCTION OF ENZYMES

Chemical transformations in biological organisms are catalysed by enzymes. Most enzymes are proteins, but nucleic acids can catalyse some reactions (for example, in RNA splicing — see p. 123), in which case the catalyst is referred to as a *ribozyme*. Compounds undergoing transformation bind to an enzyme at a site known as the *active site*, when they are said to have formed an *enzyme–substrate complex*. Folding of the polypeptide chain is important in creating the conformation giving the spatial organization of the active site.

Like all chemical reactions, *enzyme reactions are temperature-dependent*, but above about 50°C the enzyme protein starts to denature, and activity declines steeply. Extremes of pH may also affect the rate of an enzyme reaction through denaturing effects on the enzyme protein but, in addition, *variation in pH alters the enzyme activity* by its effect on the ionization of residue side-chains, especially if these are involved in the catalytic site. Thus, individual enzymes show distinctive pH–activity profiles depending on the nature of the charged groups involved.

Enzymes may be inhibited by compounds other than their substrates in a variety of ways. If the inhibitor has a structure similar to that of the substrate, it may inhibit by competing with the substrate for the active site, an effect known as *competitive inhibition*. Alternatively, the inhibitor may interact with the enzyme other than at the active site, reducing its activity through conformational or other changes, a mechanism referred to as *non-competitive inhibition*. A third major type of inhibition results when the inhibitor can only bind to the enzyme–substrate complex, which is known as *uncompetitive inhibition*. Some enzymes are subject to *allosteric modification* by compounds that bind to *allosteric sites* on the enzyme. Such enzymes are often composed of more than one subunit, and interactions between subunits play an important role in allosteric effects.

A

Enzyme properties and kinetics

Enzymes act as catalysts to accelerate the rate at which a reaction proceeds towards equilibrium.

Enzymes are proteins★. Enzymes cannot change the position of equilibrium, or the direction in which a reaction proceeds, both of which are determined by the laws of thermodynamics (see p. 70).

★While this statement is true in general terms, there are at least two examples of RNA serving as an enzyme. Ribonuclease P from *Escherichia coli* is an RNA as is the enzyme that modifies the precursor of ribosomal RNA in *Tetrahymena* (see p. 123). These findings have interesting implications for the origin of life in which it is speculated that RNA may have arisen before DNA and protein. RNA enzymes are named *ribozymes*.

B

Important variables which affect the rate of an enzyme reaction are temperature, pH and amount of enzyme present.

The 'initial rate' of an enzyme reaction is the rate at the earliest time that the reaction can be measured after mixing the reactants. It is used for all enzyme assays, and all calculations concerning enzyme kinetics. It is denoted by v_i, or more often just v.

Most enzymes start to denature above 50°C. The response to pH shows considerable variation between different enzymes. Pepsin, for example, is active at pH 1. Most enzymes show optimal activity within a few pH units either side of pH 7.

A linear relationship between the amount of enzyme and the initial rate of reaction is often (but not always) found.

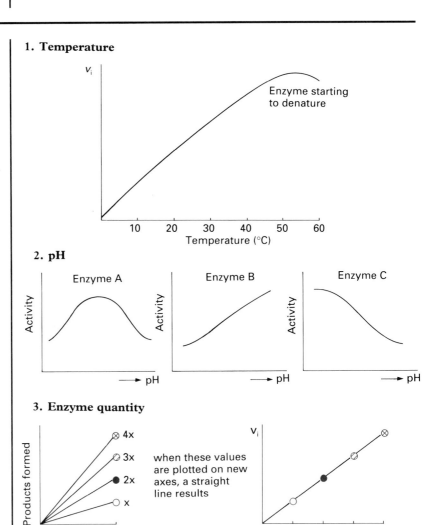

1. Temperature

Enzyme starting to denature

Temperature (°C)

2. pH

Enzyme A

Enzyme B

Enzyme C

3. Enzyme quantity

when these values are plotted on new axes, a straight line results

Time (min)

Amount of enzyme

A

The maximum rate (V_{max}) is the maximum initial rate that can theoretically occur at a given concentration of enzyme and with infinitely high substrate concentration.

The *Michaelis constant* (K_m) corresponds to the substrate concentration at which the initial rate is half the maximum rate. K_m is independent of enzyme concentration. (For a derivation of the Michaelis–Menten equation, see the following page.)

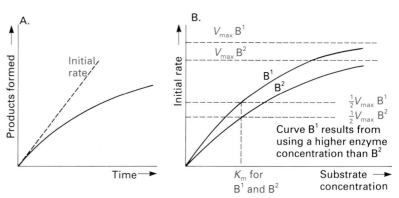

Although both of these curves look similar, B is derived from a series of measurements of initial rate as shown below.

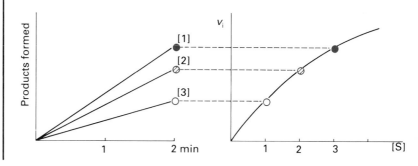

B

V_{max} and K_m are difficult to determine from plots such as those above. They are easier to determine from a plot of the reciprocal of the initial rate ($1/v$) against the reciprocal of the substrate concentration $1/[S]$, the *Lineweaver–Burk plot*.

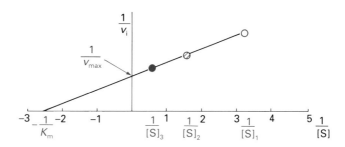

$1/v = 1/V_{max} + K_m/V_{max} \cdot 1/[S]$ is the equation for a straight line with intercept on the y-axis $1/V_{max}$ and slope $-K_m/V_{max}$. Hence the Lineweaver-Burk plot.

C

In an alternative plotting method, the Eadie–Hofstee plot, $v/[S]$ is plotted against v.

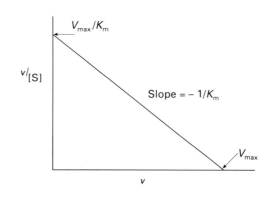

Derivation of the Michaelis–Menten equation

A mathematical analysis of the kinetics of an enzyme-catalysed reaction was proposed by Michaelis and Menten, whose names are always associated with the equation derived below. The derivation is based on a model of enzyme action in which the enzyme (E) binds to a single substrate (S) to form an enzyme–substrate complex (ES), which breaks down to form products (P) and to liberate free enzyme (E):

$$E + S \underset{k_{-1}}{\overset{k_1}{\rightleftharpoons}} ES \xrightarrow{k_2} E + P$$

k_1, k_{-1} and k_2 are rate constants for the various reactions as indicated above.

It is assumed that the reaction is in the steady state, with the concentration of ES constant, and the analysis applies only to the start of the reaction, at which point negligible amounts of products have been formed, so that the reverse reaction $E + P \rightarrow ES$ is also negligible. Then

$$\frac{d[ES]}{dt} = k_1([E_t] - [ES])[S] - k_{-1}[ES] - k_2[ES]$$

$$= 0 \text{ (} E_t \text{ is the total amount of enzyme)}$$

$$k_1[E_t][S] - k_1[ES][S] - k_{-1}[ES] - k_2[ES] = 0$$

$$k_1[E_t][S] = [ES](k_1[S] + k_{-1} + k_2)$$

$$\frac{k_1[E_t][S]}{k_1[S] + k_{-1} + k_2} = [ES]$$

Divide by k_1:

$$[ES] = \frac{[E_t][S]}{[S] + \dfrac{(k_{-1} + k_2)}{k_1}}$$

Then initial rate (v) of formation of product = $k_2[ES]$ and when the enzyme is saturated with substrate, maximum rate (V_{max}) = $k_2[E_t]$ and the Michaelis constant

$$K_m = \frac{k_{-1} + k_2}{k_1}$$

So $v = k_2[ES] = \dfrac{k_2[E_t][S]}{[S] + K_m}$

$$= \frac{V_{max}[S]}{[S] + K_m}$$

Strictly, this derivation is restricted to a single-substrate reaction. In multisubstrate reactions, if other factors are held constant and the concentration of one substrate is varied, the initial rate plotted against substrate concentration often produces a hyperbolic curve which can be analysed using the Michaelis–Menten equation.

Indeed, a hyperbolic relationship is observed when many biological events are examined kinetically. These, apart from enzyme reactions, include transport phenomena, the binding of ligands (e.g. hormones) to receptors, and drug-related effects.

k_2 is also referred to as k_{cat}. Thus, from above, $v = k_{cat}[ES]$. At very high (saturating) substrate concentrations, v approximates to V_{max}, and [ES] to [E_t], so that under these conditions k_{cat} can be calculated from $V_{max} = k_{cat}[E_t]$, if the concentration of enzyme is known.

Enzyme inhibition

Many substances inhibit enzymes and reduce the initial velocity (increase $1/v$). The Lineweaver–Burk plot reveals the mechanism of the inhibition.

Competitive inhibition

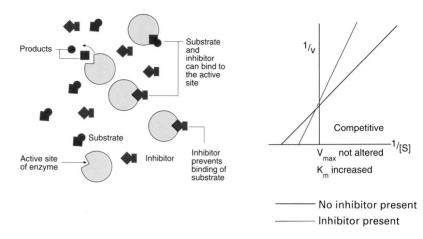

Products

Substrate and inhibitor can bind to the active site

Substrate

Active site of enzyme

Inhibitor

Inhibitor prevents binding of substrate

Competitive

V_{max} not altered

K_m increased

— No inhibitor present

— Inhibitor present

Competitive inhibition occurs when both substrate and inhibitor will fit the active site, and compete with each other to occupy it.

Non-competitive inhibition

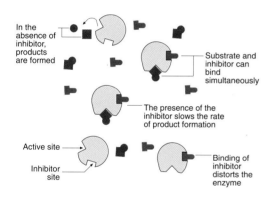

In the absence of inhibitor, products are formed

Substrate and inhibitor can bind simultaneously

The presence of the inhibitor slows the rate of product formation

Active site

Inhibitor site

Binding of inhibitor distorts the enzyme

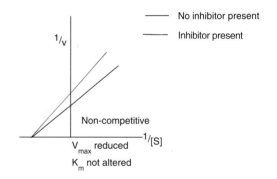

— No inhibitor present

— Inhibitor present

Non-competitive

V_{max} reduced

K_m not altered

Non-competitive inhibition occurs when the inhibitor binds to a site on the enzyme other than the active site or binds irreversibly to the active site.

Uncompetitive inhibition

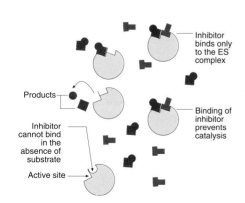

Inhibitor binds only to the ES complex

Products

Binding of inhibitor prevents catalysis

Inhibitor cannot bind in the absence of substrate

Active site

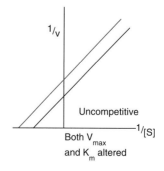

Uncompetitive

Both V_{max} and K_m altered

Uncompetitive inhibition occurs when the inhibitor binds after the substrate has bound to the enzyme, and then stops the reaction occurring.

A

International classification of enzymes

(class names, code numbers and types of reactions catalysed)

1. Oxidoreductases (oxidation–reduction reactions)
   - 1.1 Acting on CH–OH
   - 1.2 Acting on C=O
   - 1.3 Acting on C=CH–
   - 1.4 Acting on CH–NH$_2$
   - 1.5 Acting on CH–NH–
   - 1.6 Acting on NADH, NADPH

2. Transferases (transfer of functional groups)
   - 2.1 One-carbon groups
   - 2.2 Aldehydic or ketonic groups
   - 2.3 Acyl groups
   - 2.4 Glycosyl groups
   - 2.7 Phosphate groups
   - 2.8 S-containing groups

3. Hydrolases (hydrolysis reactions)
   - 3.1 Esters
   - 3.2 Glycosidic bonds
   - 3.4 Peptide bonds
   - 3.5 Other C–N bonds

4. Lyases (addition to double bonds)
   - 4.1 C=C
   - 4.2 C=O
   - 4.3 C=N–

5. Isomerases (isomerization reactions)
   - 5.1 Racemases

6. Ligases (formation of bonds with ATP cleavage)
   - 6.1 C–O
   - 6.2 C–S
   - 6.3 C–N
   - 6.4 C–C

The term 'synthase' is used for enzymes in group 4 where emphasis on the synthetic activitity of the enzyme is required.
The term 'synthetase' is now not recommended and in any event should be confined to Group 6.

B

Nature of the active site and induced fit

The substrate binds to a region of the enzyme protein known as the active site.

The active site has regions responsible for binding the substrate and regions which catalyse the reaction. There is considerable overlap between these regions.

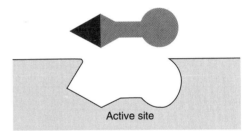

Active site

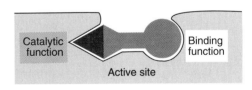

Catalytic function

Binding function

Active site

The term 'induced fit' implies that the shape of the binding site changes as a result of substrate binding, to be likened to the shape of a glove when a hand is inserted (see also p. 61B).

A

Protein folding is important in the formation of enzyme active sites.

The α-carbon backbone of *chymotrypsin* is shown as an example.

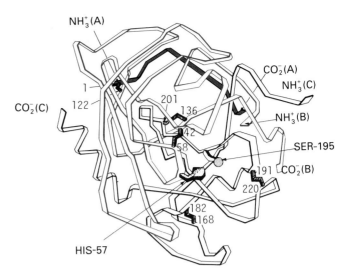

The folding of the peptide chains of the enzyme proteins is such that amino acid residues from different chains and from different regions of the same chain are brought into the correct position to form the active site, just as in haemoglobin the haem pocket is formed. Chymotrypsin is composed of three chains: A (dark red), B (light red) and C (black). His-57 is in the B-chain whilst Ser-195 is in the C-chain but both contribute to the active site. The N- and C-terminal residues are on the outside as indicated. Black bars indicate −S−S− bonds.

B

The concept that the binding of substrate induces a change in enzyme conformation is borne out by X-ray crystallography studies of hexokinase crystallized in the presence and absence of glucose. The phenomenon is known as *induced fit*.

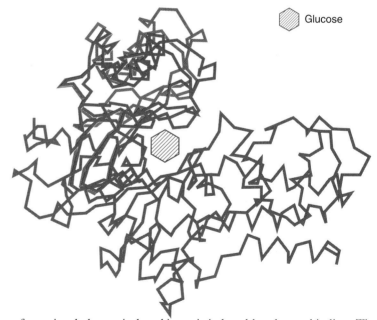

A conformational change in hexokinase is induced by glucose binding. The black lines show the α-carbon backbone of the enzyme crystallized in the presence of glucose. The red lines show the backbone of that part of the enzyme that has a different structure when crystallized in the absence of glucose.

A

The active sites of two proteolytic enzymes, chymotrypsin, shown in A, and elastase, shown in B, differ in that a bulky side-chain intrudes into the pocket in elastase. The peptide substrate is shown in red in both cases.

Conceptually, one can distinguish two functions in the active site: (1) binding and (2) catalysis. Spatially these may overlap, but the binding function often extends beyond the catalytic site (see p. 60). Thus, chymotrypsin has a larger pocket for substrate than elastase, although the catalytic sites are similar.

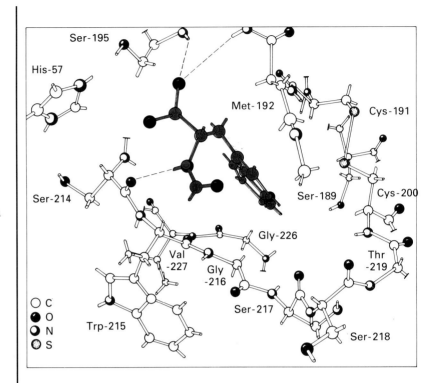

B

The substrate of elastase has a correspondingly smaller residue occupying the active site than is the case with chymotrypsin. The detailed structures reveal the precise 'tailoring' of the active site to the substrate.

Elastase, showing intrusion of the bulky side-chains of valine and threonine (in pink) at positions 216 and 226 in contrast to the glycines at these positions in chymotrypsin above.

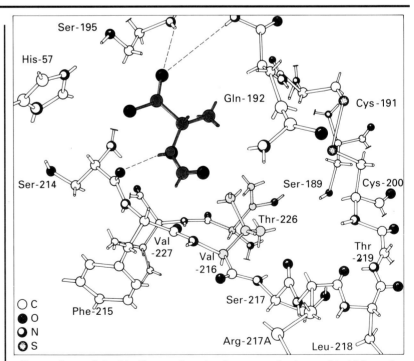

Comparison of the binding pockets in chymotrypsin (top, with N-formyl-L-tryptophan bound) and elastase (bottom, with N-formyl-L-alanine bound). The binding pocket in trypsin is very similar to that in chymotrypsin except that residue 189 is an aspartate to bind positively charged side-chains. Note the hydrogen bonds between the substrate and backbone of the enzyme.

A

In chymotrypsin and elastase, and also in a number of other proteolytic enzymes, Ser-195 plays an important role in the catalytic mechanism. Such enzymes are thus known as the *serine proteinases*.

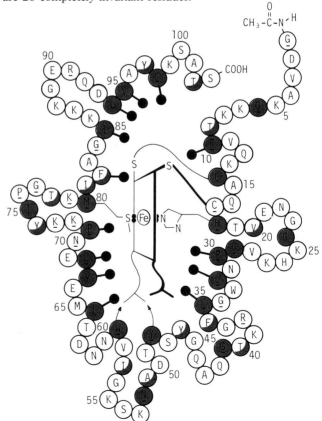

Asp-102
carboxylate

His-57
imidazole

Ser-195
hydroxyl

As shown in the structures on the previous page, His-57 lies sufficiently close to Ser-195 for one of its nitrogens to hydrogen bond to the hydroxyl of Ser-195. Shown in the figure is the fact that Asp-102 is able to hydrogen bond to the other nitrogen of His-57. This forms what is known as a *charge relay system*, in which negative charge is displaced from Asp-102 towards the hydroxyl of Ser-195, making this hydroxyl a highly reactive nucleophile, which can attack the carbonyl carbon of the peptide bond being hydrolysed. Certain nerve gases act by blocking Ser-195, and any compound (e.g. phenylmethylsulphonyl fluoride, PMSF) that blocks this group inhibits serine proteinases.

B

Evolutionary aspects of protein structure and function

To exemplify enzymes in evolution, we may trace the change in structure of cytochrome *c*.

Analysis of the structure of *homologous proteins* (see p. 65B for definition), i.e. those having an identical function, from many species that have evolved over widely different evolutionary pathways indicates to what extent a structure has been *conserved* during evolution.

The amino acid sequence of cytochrome *c* from more than 60 different species is known. These sequences vary considerably at certain positions. However, every cytochrome *c* so far sequenced has histidine at a position corresponding to 18 in the human; that is to say, it is invariant. Altogether there are 26 completely invariant residues.

Chain-packing diagram of tuna reduced cytochrome *c*, showing evolutionary invariant amino acids (underlined capitals), buried side-chains (red circles), side-chains packed against the haem (black dots) and side-chains half buried (half-red circles).

Superfamilies: the serpins, homologies and motifs

The relative ease of sequencing proteins and peptides using automatic sequencing instruments and cDNA technology has led to the installation of extensive databanks of protein sequences. From these it has been possible to deduce certain patterns common to proteins of similar function and evolutionary origin.
For explanation of arrows and coils see 65A

Groups of proteins related by structure and function are referred to as *superfamilies*. An example of one such superfamily is given by a group of proteins known as the *serpins* (*ser*ine *p*roteinase *in*hibitors). Prominent members of this family are two plasma proteins, α_1-antitrypsin, which inhibits elastase (see p. 35), and antithrombin, which is an inhibitor of thrombin, a protein that among other actions promotes blood coagulation. The structure of α_1-antitrypsin is shown below.

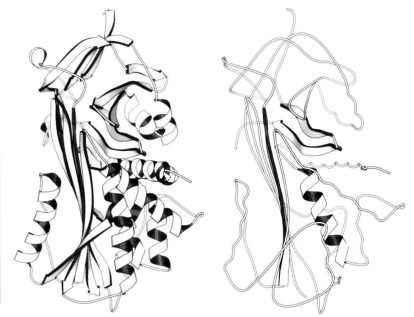

The left-hand figure shows a ribbon diagram of the complete molecule of α_1-antitrypsin. On the right, those structures exhibiting the highest degree of homology with other members of the superfamily are picked out. The helix shown is helix B, the upper of the two short β-sheets is 4B, the lower is 5B and the long vertical sheet is 3A. The actual amino acid sequences of these regions are shown below, with the conserved residues in heavy type.

	Helix B	Sheet 3A	Sheet 4B	Sheet 5B
Human antitrypsin	-PVSIATAFAMLSLGT-	-LVNYIFFKGKW-	-KPFVFLMI-	-FMGKVVNPT-
Heparin cofactor II	-PVGISTAMGMISLGL-	-INLCIYFKGSW-	-RPFLFLIY-	-FMGRVANPS-
Antithrombin	-PLSISTAFAMTKLGA-	-LVNTIYFKGLW-	-RPFLVFIR-	-FMGRVANPC-
C1 inhibitor	-PFSIASLLTQVLLGA-	-LVNAIYLSAKW-	-QPFLFVLW-	-FMGRVYDPR-
Protein C inhibitor	-PVSISMALAMLSLGT-	-LVNYIFFKGTW-	-QPFIIMIF-	-FLARVMNPV-
Corticosteroid BG	-PVSISMALAMLSLGA-	-MVNYIFFKAKW-	-RPFLMFIV-	-FLGKVNRP

The serpins are thought to have developed by divergent evolution over a period of some 500 million years. The original ancestral protein is presumed to have had the function of a serine proteinase inhibitor, but some members of the superfamily, currently included on the grounds of homology, have lost this function and developed specialized roles, e.g. corticosteroid-binding globulin (corticosteroid BG — see above), or have no recognized function, such as ovalbumin. Others retain this function, including the inhibitor of C1 (the first component of complement) and protein C (a normal plasma serine proteinase that is activated by thrombin and cleaves clotting factors Va and VIIIa (see p. 244C) to inactive forms). Hard and fast rules do not exist for allocating proteins to superfamilies, and this is done by general consensus. Convincing evidence of similarity of function is sought, in addition to the presence of highly conserved amino acid residues, although the latter consideration predominates, as in the case of ovalbumin, the role of which in the hen's egg, apart from being a nutritional reserve, remains obscure.

A

The *dehydrogenase structures* are an excellent example of the use of pleated sheet, helical and random coil regions in a globular protein.

Pleated sheets are represented by the broad arrows. The cylinders represent α-helical regions. The thin tubular portions indicate regions of random coil (i.e. no defined secondary structure). Numbers without prefixes refer to amino acid positions. Prefixes α and β label helical and pleated sheet regions, respectively.

α-Helical regions are sometimes represented by coiled ribbons as shown on page 64.

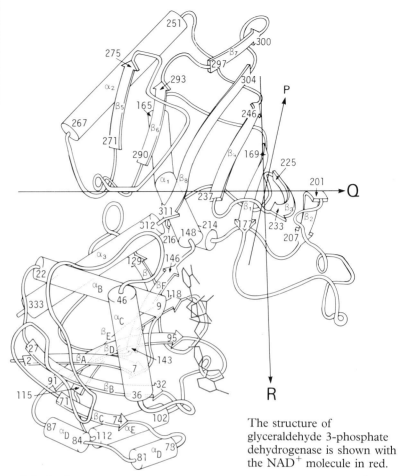

The structure of glyceraldehyde 3-phosphate dehydrogenase is shown with the NAD$^+$ molecule in red.

A series of β-sheets (β$_A$, β$_B$, β$_C$, β$_D$, β$_E$ and β$_F$, in the structure shown) is characteristically involved in NAD$^+$ binding in a number of dehydrogenases. The hydrogen bonding in the binding sites and charge relay system described on the previous two pages is found in a number of serine proteinases that have virtually no other sequence similarities. Whether these structures have evolved by selection of the essential features of the active site rather than conservation of residues during divergence from a common precursor is a controversial question, but has given rise to the term *convergent evolution*.

B

Homologous proteins.

Having determined the primary structure of a protein either by direct methods or by deduction from the nucleotide sequence of the gene, it is common to use a computer database to see whether the new structure bears any similarity to that of a previously determined structure. By this means it may be possible to identify a *consensus sequence* of amino acid residues which represent a functional domain, the biological role of which has already been identified. If there is a considerable (more than 30%) similarity between the structures it has been common to state that the proteins are *homologous*. There are others who would confine such a term to those proteins which have been shown from both their primary structure and organization of their genes to have been derived by the duplication of a common ancestral gene.

Two terms that are useful for describing the structure within the polypeptide chains are *domain* and *motif*. A domain is defined as a polypeptide or part of a chain that can be independently folded into a stable tertiary structure. A domain may also have a functional role, e.g. binding to DNA or a hormone. A motif is a combination of a few secondary structural elements with a specific geometric arrangement. Again, some can be associated with a particular function. Domains are built from structural motifs.

A

Allosteric properties of enzymes

The rate of an enzyme reaction can be affected by the presence of a substance (such as cyclic AMP) that combines with the enzyme protein at a site other than the active site. Such a substance is known as an *allosteric modulator, modifier* or *effector*, and its binding site as an *allosteric site.* Cooperativity is seen in binding of substrate, similar to that found in the binding of O_2 to haemoglobin. Characteristically, such enzymes exhibit sigmoid v versus *[S]* curves.

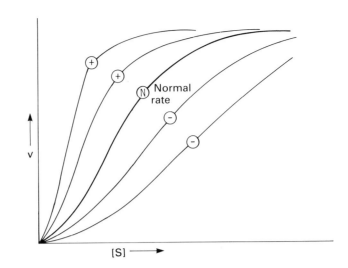

Influence of modulators on the substrate (S)–velocity (v) curves of a regulatory enzyme. *Positive modulation* (+), *negative modulation* (−), in each case at two different concentrations of modulator, the effect increasing with increase in modulator concentration.

B

The Monod–Wyman–Changeux model.

The behaviour of an allosteric molecule may be analysed using this model.

The model envisages the binding of substrate to one of two different states, the T (tense) state and the R (relaxed) state (compare haemoglobin — see p. 23). The R-state predominates when an allosteric modifier is bound. The T-state is the predominant form when the protein is unligated. Cooperativity, as evidenced by an enzyme in sigmoid v versus *[S]* plots is exhibited only by the T-state. If a molecule is entirely in the R-state as a result of binding an allosteric activator, non-sigmoid (i.e. Michaelis–Menten) kinetics result. Intermediate states may be thought of as involving hybrid forms, in which some subunits with substrate bound may be in the R-state, while some subunits to which no substrate is bound can be in the T-state.

A

Regulation of enzyme activity

Protein kinases, cyclic AMP and allosteric effects

The phosphorylation of enzyme proteins, and other proteins, is an important regulatory mechanism. The phosphorylated enzyme may be *more* active or *less* active, depending on the enzyme.

In the case of kinases involved in regulation of pathways of intermediary metabolism the phosphate is esterified to hydroxyls of serine or threonine residues. Other protein kinases, involved in growth regulation, add phosphate to the phenolic group of tyrosine (see p. 272).

B

Cyclic adenosine 3',5'–monophosphate (cyclic AMP, cAMP) is an important *allosteric regulator* of certain protein kinases.

cAMP is known as a *second messenger* (the hormone being the first).

The enzyme *adenylylcyclase* converts ATP to cAMP (see also p. 264):

$$ATP \rightarrow cAMP + PP_i\star$$

Phosphodiesterase hydrolyses the bond between the phosphate and the 3' hydroxyl of the ribose. Between them, adenylyl cyclase and phosphodiesterase thus regulate the concentration of cAMP.

★PP_i is pyrophosphate (P_i is the symbol for inorganic phosphate).

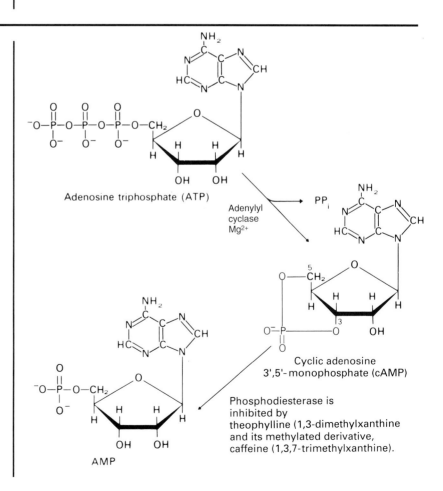

Adenosine triphosphate (ATP)

Adenylyl cyclase Mg^{2+}

Cyclic adenosine 3',5'-monophosphate (cAMP)

Phosphodiesterase is inhibited by theophylline (1,3-dimethylxanthine and its methylated derivative, caffeine (1,3,7-trimethylxanthine).

AMP

A

Three-dimensional representations of some allosteric enzymes have been deduced. One example of such an allosterically regulated enzyme is *phosphofructokinase* (PFK). This enzyme is inhibited by one of its own substrates (ATP).

The different sites indicated are for binding substrates and effectors as follows: A, F6P; C*, ADP activator; A', FBP; C', ATP inhibitor; B, ATP substrate.

*Not shown.

Most allosteric enzymes are oligomeric proteins and, as outlined on p. 66, this is one basis for explaining the sigmoid kinetic curve. Interaction between subunits is important in cooperative effects. Rabbit muscle PFK is composed of subunits, each of which consists of two pear-shaped domains linked by a helical region.

Four of these subunits pack together in the mammalian enzyme. Such a tetramer carries four active sites, and a number of allosteric sites. The active site is in a cleft *on* each subunit as shown by B. The allosteric sites are thought to lie in clefts formed *between* subunits, as shown by C' at the bottom and A and A' in the clefts at the top.

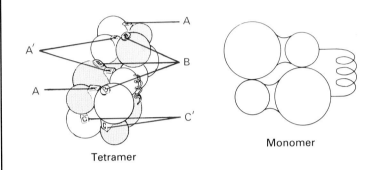

Tetramer

Monomer

B

A more detailed diagram indicates more clearly the nature of the allosteric sites in PFK, in this example from *Bacillus stearothermophilus*.

This has smaller subunits than the rabbit muscle enzyme described above, but the diagram illustrates the way in which allosteric sites lie between subunits. (Each of the monomers of the rabbit muscle enzyme corresponds to a dimer of the bacterial enzyme, which is tetrameric, so that each dimer of the mammalian enzyme corresponds to the bacterial tetramer.)

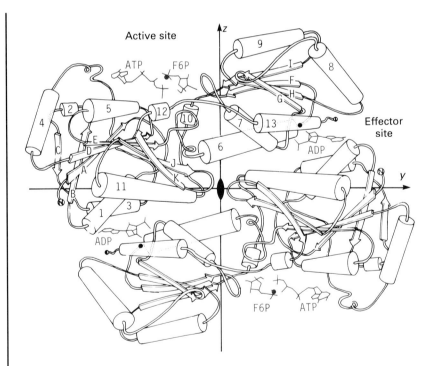

A schematic view of two subunits of the PFK tetramer, viewed along the x-axis. β-sheet strands are represented by arrows (A–K) and α-helices by cylinders (1–13). Each subunit consists of two domains: domain 1 is on the left in the upper subunit. The substrates ATP and F6P are shown in the active site, and the activator ADP in the effector site.

A

In the case of phosphorylase, the enzyme responsible for degrading glycogen (see p. 199A, B), it is known in detail how the structure changes on the transition from the T- to the R-state.

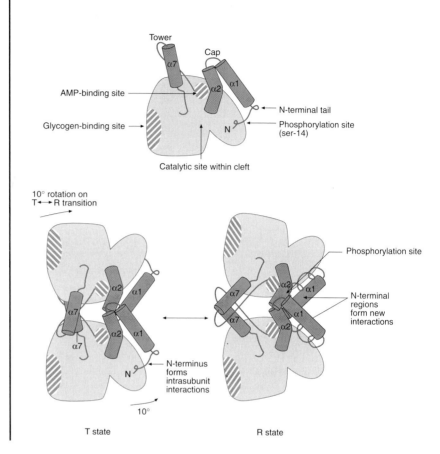

The active enzyme exists as dimers. In the upper part of the diagram, a monomer is represented to show some of the main features of the molecule. Three of the helices ($\alpha 1$, $\alpha 2$ and $\alpha 7$, which is known as the tower helix) are shown, with the rest of the molecule represented by the outline shape, indicating binding sites, the target site of the serine kinase (phosphorylase kinase), and the catalytic site. Residues 40–45 are referred to as the 'cap'. In the lower part of the diagram, dimers are shown, indicating the changes that occur as a result of the T → R transition. As indicated, the monomers rotate $10°$ in relation to the vertical axis, and the three helices change their orientation, especially in the case of $\alpha 7$. Movement of the tower helix indirectly leads to a replacement of an aspartate by an arginine residue at the catalytic site. When Ser-14 is phosphorylated, the 20 N-terminal residues form an ordered helical conformation, also found in the dephospho R-state. See p. 199 for further discussion.

B

Protein phosphatases

In any regulatory system, a mechanism must exist for reversing the regulatory effect. Thus, in the protein phosphorylation system, there are not only kinases, but phosphatases.

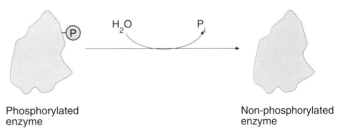

Phosphorylated enzyme

Non-phosphorylated enzyme

Several protein phosphatases can dephosphorylate glycogen phosphorylase. One of these, designated protein phosphatase 1, is ATP- and Mg^{2+}-dependent. Another, protein phosphatase 2B in the current nomenclature, also called calcineurin, is Ca^{2+}/calmodulin-dependent.

A

Energy, metabolic control and equilibria

The laws of thermodynamics determine the course of enzyme reactions.

Associated with every compound is a quantity, G, known as its *free energy*.

For a molecule in solution, free energy is concentration-dependent. The free energy of a particular component of a solution is equal to the free energy per mole times the number of moles of that component. *The standard free energy* (G°) *is equal to the free energy of the component at unit activity* (in dilute solution this is approximately 1 M). The molar free energy of a particular component of a solution is often referred to as the *chemical potential* of that component.

Thus, for the ith component of a solution

$$G_i = G_i{}^\circ + RT \ln a_i \dots\dots\dots (1)$$

where R is the gas constant (8.134 J mol^{-1} K^{-1}), T the absolute temperature, and a_i the activity of the ith component (in dilute solution, activity approximates to concentration).

B

For a chemical reaction there is a quantity, ΔG, known as the *free energy change* of the reaction, which is calculated as the difference between the free energies of the reactants and products at a particular instant in time. At equilibrium $\Delta G = 0$.

For a reaction

$$A + B \xrightarrow{\longleftarrow} C + D$$

then, at any instant in time,

$$\Delta G = G_C + G_D - G_A - G_B$$

Using equation (1), and observing that $G_C^\circ + G_D^\circ - G_A^\circ - G_B^\circ$ is termed the *standard free energy change* (ΔG°) of the reaction, then

$$\Delta G = \Delta G^\circ + RT \ln \frac{[C]_C\,[D]_D}{[A]_A\,[B]_B} \dots\dots\dots (2)$$

Where $[A]_A$, $[B]_B$, $[C]_C$ and $[D]_D$ are the concentrations pertaining at the moment for which ΔG is calculated.

The magnitude and sign of ΔG determine the direction in which the reaction will proceed. It will always proceed in a direction that tends to bring the free energies of the components on one side of the equation equal to those on the other. When the free energies on both sides of the reaction are equal, equilibrium has been reached. It follows that at equilibrium the free energy change is zero. If ΔG is positive, ΔG for the reaction in the reverse direction will be negative, and thus the reaction will proceed in the reverse direction.

C

The standard free energy change (i.e. when all reactants are at unit activity) is the quantity which is used to compare free energy change between one reaction and another, and is a constant at constant temperature.

The standard free energy change is the free energy change when [A], [B], [C] and [D] represent unit activity (for practical purposes 1 M) since under these conditions, the second term on the right of equation (2) above becomes zero (ln 1 = 0) and $\Delta G = \Delta G^\circ$.

If we wish to evaluate ΔG°, we can do this by writing an equation for the equilibrium position, because this is the point at which $\triangle G = 0$. Thus, at equilibrium,

$$0 = \Delta G^\circ + RT \ln \frac{[C]_E\,[D]_E}{[A]_E\,[B]_E}$$

or

$$\Delta G^{\circ\prime} = -RT \ln K_{eq}$$

where $[A]_E$, $[B]_E$, $[C]_E$ and $[D]_E$ are equilibrium concentrations of A, B, C and D.

If the equilibrium constant is known, therefore, the standard free energy change can be readily found.

Note: $\Delta G^{\circ\prime}$ is used to denote that one component, H$^+$ is not at unit activity, when measurements are made at pH 7.0.

A

Free energy changes are involved in the reactions of biochemical pathways, which proceed in the direction of an overall negative free energy change.

It is often pointed out that the standard free energy change for the hydrolysis of ATP is negative and large compared with that for a phosphate ester, such as glycerol 3-phosphate. This in the past has given rise to the expression 'high-energy compound'.

$$ATP^{4-} + H_2O \longrightarrow ADP^{3-} + P_i + H^+ \quad \Delta G^{\circ\prime} = -30.5 \text{ kJ mol}^{-1}$$
$$\text{Glycerol 3-phosphate} + H_2O \longrightarrow \text{Glycerol} + P_i \quad \Delta G^{\circ\prime} = -9.2 \text{ kJ mol}^{-1}$$

For this reason, reactions involving ATP often proceed in the direction that results in formation of ADP and P_i. However, in order that ATP stores may be replenished, the reverse reaction must also occur in biological systems. This will be the case when the reaction also involves other compounds of such a nature that the reaction results in a large negative free energy change, again dependent on the concentrations of reactants and products.

The most important example of this is when phosphorylation of ADP is coupled to the oxidation of substrates by the electron transport chain (see p. 185A).

$$NADH + H^+ + O_2 \longrightarrow NAD^+ + H_2O \quad \Delta G^{\circ\prime} = -222 \text{ kJ mol}^{-1}$$

Another example is

$$\text{Phosphoenolpyruvate} + ADP \longrightarrow \text{Pyruvate} + ATP$$

(The standard free energy change for the hydrolysis of phosphoenolpyruvate is -62 kJ mol^{-1}.)

B

Regulation of metabolic pathways

The direction of any chemical reaction, such as an enzymic reaction, is regulated by the forementioned thermodynamic laws, which include a component for the concentrations of the reactants. These concentrations will depend on the rate of the various enzymic reactions.

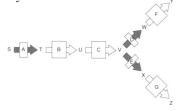

Regulation often occurs at the beginning of pathways or at branch points as shown above.

Individual pathways can be regulated by allosteric inhibition (or activation) by certain metabolites of the same or other pathways.

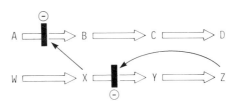

In the pathway A → B → C, intermediate B will tend to accumulate if its formation from A is faster than its conversion to C. At the same time, its rate of conversion to C will in part be regulated by its concentration. The rate at which a pathway proceeds is thus constantly regulated by the supply of substrate, and the rates of the individual enzymes. These rates will be determined partly by the amount of enzyme protein, and partly by the presence of activators and inhibitors. An important consequence of the phenomenon of allosteric regulation is that intermediates from pathways can regulate enzymes for chemically unrelated reactions, if an enzyme has an allosteric site for the intermediate in question. Thus, regulation can occur between one pathway and another, as indicated in the diagram, or by pathway products regulating earlier enzymes of the same pathway. This latter effect is referred to as *feedback regulation*, or, if appropriate, *feedback inhibition*.

More widespread and general regulation is exerted by changes in ATP/ADP, NADH/NAD$^+$ or NADPH/NADP$^+$ ratios. For example, changes in the NADH/NAD$^+$ ratio will affect all dehydrogenases utilizing this pair of coenzymes in a given cell compartment.

A

The concept of equilibrium yields an equation of fundamental importance in the analysis of reactions between two or more molecules.

The equilibrium equation, $[A][B]/[C][D] = K_{eq}$, provides a basis for the analysis of many diverse but conceptually related phenomena, including acid–base behaviour, and the binding of ligands (hormones to receptors, allosteric modifiers to enzymes and other cooperative phenomena).

B

One of the most everyday relationships that can be derived from the equilibrium equation concerns pH. The *Henderson–Hasselbalch equation* is used to relate pH to the concentration of acid and base in a buffer.

In general, for an equilibrium $[AB] \rightleftharpoons [A] + [B]$,

$$\frac{[AB]}{[A][B]} = K_{association} \quad \text{and} \quad \frac{[A][B]}{[AB]} = K_{dissociation}$$

For pH equilibria, HA is a conjugate acid and A^- a conjugate base

$$\frac{[H^+][A^-]}{[HA]} = K_a \tag{1}$$

(K_a is the acid dissociation constant).

Taking logs,

$$\log [H^+] + \log [A^-] - \log[HA] = \log K_a \tag{2}$$

pH is defined as $-\log[H^+]$, pK_a as $-\log K_a$

Thus, multiplying equation (2) by -1 and rearranging,

$$pH = pK_a + \log \frac{[A^-]}{[HA]} \tag{3}$$

This is the Henderson–Hasselbalch equation.

C

Analysis of ligand binding

It is a characteristic of binding phenomena that ligands exhibit saturation kinetics.

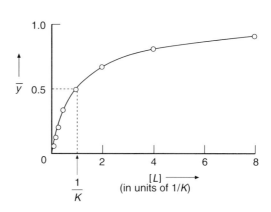

If, in the binding of a ligand (L) to its receptor (R), [L] is the concentration of free ligand, and [RL] is the concentration of bound ligand, then the ratio of [RL] to all forms (bound or free) of the receptor, i.e. $[RL]/[R_{Tot}]$, is known as the saturation fraction, and is given the symbol y. A plot of y against [L] is referred to as the *adsorption isotherm*. K = association constant of the complex.

A

Scatchard analysis

The quantities we most often need to extract from binding data concern (1) the dissociation constant for the complex and (2) the total number of receptor molecules. This can be obtained from the *Scatchard plot.*

The association constant (K_a) = [RL]/[L][R], where [L] and [RL] are as defined previously, and [R] is the concentration of unbound receptor. Since

$$[R] = [R_{Tot}] - [RL]$$

then

$$[RL]/([L][R_{Tot}] - [L][RL]) = K_a$$

This can be readily re-arranged to give

$$[RL]/[L] = - K_a[RL] + K_a[R_{Tot}]$$

which is the equation of a straight line with negative slope K_a and intercept on the abscissa $[R_{Tot}]$.

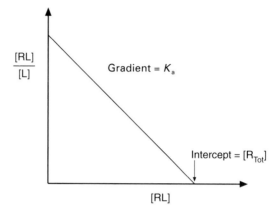

Gradient = K_a

Intercept = $[R_{Tot}]$

Alternative Scatchard plots can be made. For example, a plot of $y/[L_{Tot}]$ versus $[L_{Tot}]$ gives a plot of similar appearance, with the intercept on the ordinate yielding K_0. This plot is reminiscent of the Eadie–Hofstee plot (see p. 57C), where v is proportional to the saturation of the enzyme with substrate, and thus analogous to y. However, the method of plotting illustrated above is almost invariably employed, because using ligands labelled with radioactivity or fluorescence [RL] and [L] can readily be ascertained experimentally.

B

A linear Scatchard plot is obtained only where there is one binding constant. If sites with different binding constants are involved, a plot which is biphasic or more complex results, from which the number of sites of different types can be determined.

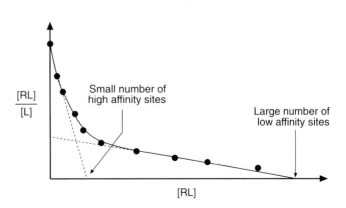

Small number of high affinity sites

Large number of low affinity sites

Cooperative phenomena — the Hill plot

Cooperativity can best be analysed using the Hill equation. This yields the Hill plot, from which can be determined the *Hill coefficient*, a measure of cooperativity.

The Hill equation is used to analyse binding situations where cooperativity is manifested, that is, where binding sites exist for more ligands than one and binding of each ligand facilitates binding of successive ligands. The simplifying assumption is made that the ligand fully occupies each successive site before any of the next site is occupied. For example, for a molecule with four binding sites, such as haemoglobin, only five states are envisaged as in the figure.

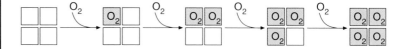

An equilibrium expression can be written as

$$K = [R][L]^n/[RL_n]$$

where $[RL_n]$ is the concentration of receptor with n ligands bound.

In such a case,

$$y = [RL_n]/[R_{Tot}]$$

Since

$$[RL_n] = [R_{Tot}] - [R]$$

then

$$1 - y = [R]/[R_{Tot}]$$

and we can obtain the equation

$$y/(1 - y) = [RL_n]/[R]$$

But

$$[RL_n] = [R][L]^n/K$$

and so

$$y/(1 - y) = [L]^n/K$$

giving

$$\log y/(1 - y) = n \log[L] - \log K$$

the Hill equation. n is known as the *Hill coefficient*.

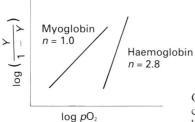

Comparison of n with the number of binding sites gives a measure of cooperativity, a ratio of 1 indicating complete cooperativity. For haemoglobin (see plot), this value is 0.7 (2.8/4).

A

Multiple forms of enzymes, isoenzymes

Isoenzymes are proteins with different structures which catalyse identical reactions (also called isozymes).

In the case of LDH M, α-hydroxybutyrate is the preferred substrate, so it is referred to as α-hydroxybutyrate dehydrogenase (αHBD) even though α-hydroxybutyrate does not occur in the body.

Lactate dehydrogenases (LDH)

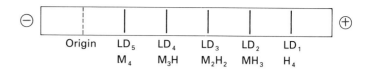

After electrophoresis of a tissue extract, several protein bands possessing LDH activity may be detected. This arises because two different subunits of different charge are present in the cell. These are termed M (muscle) and H (heart). Permutations of these subunit types among the four subunits of the active enzyme give five possible combinations.

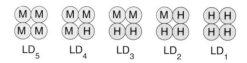

Although isoenzymes catalyse the same reaction the affinity of the enzyme for the substrate may be different. Thus, hexokinase I and hexokinase IV (glucokinase) in liver have markedly different K_m values for glucose, but both catalyse the reaction

$$Glucose + ATP \rightarrow Glucose\ 6\text{-phosphate} + ADP$$

B

Multiple forms of enzymes can arise for a number of different reasons.

Group	Reason of multiplicity	Example
1.	Genetically independent proteins★	Malate dehydrogenase in mitochondria and cytosol
2.	Heteropolymers (hybrids) of two or more polypeptide chains, noncovalently bound★	Hybrid forms of lactate dehydrogenase
3.	Genetic variants (allelozymes)★	Glucose 6-phosphate dehydrogenases in man
4.	Conjugated or derived proteins a. Proteins conjugated with other groups b. Proteins derived from single polypeptide chains	Phosphorylase a, glycogen synthase D The family of chymotrypsins arising from chymotrypsinogen
5.	Polymers of a single subunit	Glutamate dehydrogenase of M_r 1 000 000 and 250 000
6.	Conformationally different forms	All allosteric modifications of enzymes

★ These classes fall into the category of isoenzymes.

A

Enzymes in diagnosis

The assay of serum enzymes is used as an important aid to diagnosis. The level of activity of a number of enzymes is raised in different pathological conditions.

It is often useful to measure the level of activity at more than one time, as the time course over which the activity of each enzyme changes is characteristic of the condition.

The diagnosis of *viral hepatitis.*

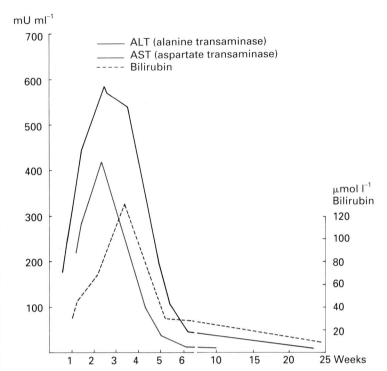

In cases of acute viral hepatitis there is a rapid rise in the level of serum transaminase activity and this occurs considerably before there is jaundice, shown by a rise in the serum bilirubin level.

B

The diagnosis of *myocardial infarction.*

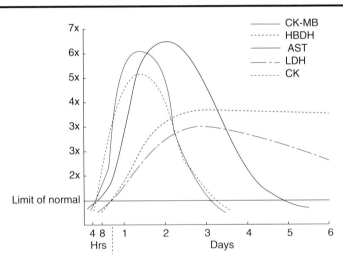

Aspartate transaminase (AST) and creatine kinase (CK) rise about 6 hours after the acute episode, 3-hydroxybutyrate dehydrogenase (HBDH) and lactate dehydrogenase (LDH) much later. Three isoenzymes of CK have been identified, with subunits denoted M (for muscle) or B (for brain). CK-MM is the predominant form in skeletal muscle, which also contains about 3% CK-MB, and CK-BB is found in cerebral tissue. More than 40% of the total CK activity is found as CK-MB in heart muscle, which is thus often measured in cases of myocardial infarction as a more precise indicator of heart muscle damage.

A

Some enzymes and proteins of diagnostic interest

Enzyme	Principal conditions in which level of activity in serum is elevated
Amylase	Acute pancreatitis
Acid phosphatase (optimum pH 5)	Prostatic carcinoma
Alkaline phosphatase (optimum pH 10)	Diagnosis of liver disease, especially biliary obstruction and detection of osteoblastic bone disease, e.g. rickets
Aspartate transaminase (AST—previously GOT)	Myocardial infarction. Liver disease, especially with liver cell damage
Alanine transaminase (ALT—previously GPT)	Liver disease, especially with liver cell damage
Lactate dehydrogenase (LDH)	Myocardial infarction, but also increased in many other diseases (liver disease, some blood diseases)
Creatine kinase (CK)	Myocardial infarction and skeletal muscle disease (muscular dystrophy, dermatomyositis)
γ-Glutamyl transferase (γGT)	Diagnosis of liver disease, particularly biliary obstruction, and alcoholism.

γ-Glutamyl transferase (γGT) catalyses the reaction

Amino acid + Glutathione $\xrightarrow{\gamma GT}$ γ-Glutamyl amino acid + Cysteinylglycine

It may be involved in amino acid transport across membranes or in the salvage of glutathione. It is elevated in the plasma of alcoholics, and also of epileptics taking barbiturates. This is because consumption of alcohol or barbiturates causes considerable proliferation of the endoplasmic reticulum, inducing high levels of the enzyme.

α-Fetoprotein
A protein formed in fetal liver, and found in the serum in adults with hepatoma and teratoma. Assay in the amniotic fluid aids in the diagnosis of neural tube defects and Down's syndrome.

Enzyme in urine
N-Acetylglucosaminidase in the urine can be used to indicate renal transplant rejection.

B

Diagnostic enzyme assay.

Measurement of the rate of formation of products gives an indication of the activity of an enzyme, but not necessarily of the amount of enzyme protein.

In many cases where an enzyme is measured in the blood, the interest is in determining the amount of enzyme protein. Since assay of an enzyme can be affected by the presence of activators or inhibitors, the activity does not always relate precisely to the amount of enzyme protein. This can be ascertained by other methods, such as radioimmunoassay (see p. 47). On the other hand, when enzymes are assayed in tissues where a genetic abnormality is suspected, it is the activity that is important, since a mutant enzyme may be produced in normal amounts, but its activity may be lacking. A *coupled* assay is one in which a second enzyme is used to act on the product of the enzyme of primary interest, often using a second enzyme that has NAD^+ as coenzyme, so that the rate can be followed by measuring formation of NADH, which can be done very conveniently at 340 nm.

A

Metabolism of ethanol

Formation of acetaldehyde and acetate.

Ethanol is metabolized in the human through the action of two enzymes:

Liver cytosolic alcohol dehydrogenase (ADH)

$$C_2H_5OH + NAD^+ \rightleftarrows CH_3CHO + NADH$$

Acetaldehyde dehydrogenase (ALDH)

$$CH_3CHO + H_2O + NAD^+ \rightleftarrows CH_3COOH + NADH + H^+$$

Acetate is converted to acetyl coenzyme A.

Many of the unpleasant side-effects of excess ethanol consumption are due to the formation of acetaldehyde.

B

The role of the isoenzymes of the dehydrogenases of ethanol and acetaldehyde.

You should be careful when you treat your oriental friend—for they may flush and experience nausea and palpitations in response to even low amounts of ethanol.

Alcohol dehydrogenase

Human ADH is a dimer and is polygenie, being formed from three different subunits, α, β and γ. The Caucasian β_1 subunit is substituted in a very high proportion (90%) of orientals by β_2. On starch gel electrophoresis we find:

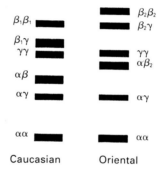

Caucasian   Oriental

Kinetic properties of alcohol dehydrogenase isoenzymes

Composition of dimer	$\beta_1\beta_1$		$\beta_2\beta_2$	
Activity	ph 8.5	pH 10.0	pH 8.5	pH 10.0
Units mg^{-1}	0.39	0.80	32	5.0

It can be seen that at physiological pH the $\beta_2\beta_2$ enzyme is much more active than the $\beta_1\beta_1$ and hence much more acetaldehyde is formed. The structural difference between β_1 and β_2 resides in a change at the coenzyme-binding site of His for Arg.

Acetaldehyde dehydrogenase

Caucasians have two isoenzymes for the oxidation of acetaldehyde, ALDH–1 and ALDH–2, while about 40% of orientals have only ALDH–1. Since ALDH–2, the mitochondrial enzyme, has a high affinity for its substrate, its absence will increase the concentration of acetaldehyde in atypical subjects. The absence of ALDH–2 is probably the main cause of the undesirable effect of ethanol on orientals. A number of ALDH inhibitors have been devised, of which disulfiram (Antabuse) is the best known, and these have been useful for the treatment of alcoholics in that they are converted into flushers.

NUCLEIC ACIDS AND PROTEIN BIOSYNTHESIS

This chapter considers: (1) the strategy of protein biosynthesis bearing in mind what has been learnt about protein structure; (2) the structure and biosynthesis of the nucleic acids, DNA and RNA, and the events that take place in the cell (the action of drugs on these processes is also described); (3) the biosynthesis of proteins by translation of messenger RNA and the role of transfer RNA and ribosomal RNA and the inhibition of these events by drugs; (4) the control of protein synthesis; (5) molecular cell biology; and (6) recombinant DNA and its use in the diagnosis of disease.

1. STRATEGIES FOR PROTEIN BIOSYNTHESIS

A

Dynamic state of body proteins

The metabolic state of the body proteins is an example of the dynamic state of the body constituents.

Proteins are synthesized from free amino acids; there are no free peptide intermediates.

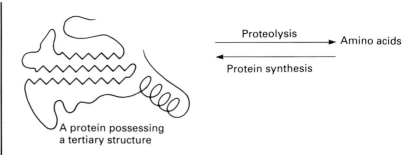

A protein possessing
a tertiary structure

The body obviously needs to synthesize protein to provide a net increase for (1) growth, (2) the replacement of cells after their death and (3) milk proteins during lactation. However, even in an adult all the proteins, with the exception of collagen, are in a constant state of degradation and resynthesis so that the total amount of protein synthesized in the body is much greater than the net increase. An adult man synthesizes about 400 g of protein each day. The rate of turnover of a protein is measured as the *half-life*, i.e. the time taken for half the molecules to be degraded to amino acids, usually measured in a few days.

B

Specialization among the various tissues

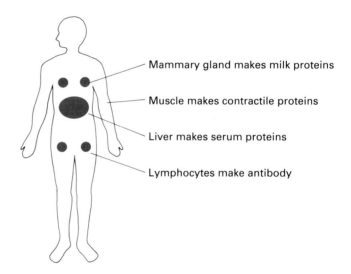

Mammary gland makes milk proteins

Muscle makes contractile proteins

Liver makes serum proteins

Lymphocytes make antibody

The protein-synthesizing activity of the body is subdivided among the various tissues and their component cells. Often the synthesis of a particular protein is confined uniquely to a cell type, e.g. serum albumin to the parenchymal cells of the liver, or the B-lymphocytes for the synthesis of immunoglobulin G. Sometimes tumours possess the ability to synthesize proteins other than those expected of the original cell, a phenomenon known as ectopic protein synthesis, e.g. the synthesis of α-fetoprotein (typical of the embryo) by hepatoma cells. The detection of serum α-fetoprotein is useful in the diagnosis of neural tube defects, Down's syndrome and primary liver cancer (see p. 77).

A

Incorporation of amino acids

The polypeptide chain of proteins is assembled from 20 different amino acids.

Some of the amino acid residues are subsequently modified. This is known as post-translational modification.

The following amino acids participate in the formation of polypeptide chains:

Glycine	Proline	*Asparagine*
Valine	Tryptophan	Lysine
Alanine	Tyrosine	Arginine
Leucine	Phenylalanine	Histidine
Isoleucine	Glutamic acid	Methionine
Serine	*Glutamine*	Cysteine
Threonine	Aspartic acid	

Note the presence of *glutamine* and *asparagine*.
Note the absence of *hydroxyproline* and *hydroxylysine,* and of *phosphoserine* and *phosphothreonine.* These modified amino acids are important constituents of collagen and complement, and of the phosphoproteins, respectively. The parent amino acids are modified by hydroxylation or phosphorylation after incorporation into the polypeptide chains. Cystine is formed from two cysteine residues. More recently, *phosphotyrosine* has been detected, particularly in proteins involved in the mediation of receptor action.

B

Multichain proteins

Multichain proteins may be assembled from individually synthesized chains or by limited proteolytic cleavage of larger proteins (e.g. insulin—see p. 82).

The formation of the correct S–S bonds is assisted by an enzyme, protein disulphide isomerase (see p. 19). The interaction of the separate chains may be assisted by other proteins known as chaperones (see p. 19), as in the case of the immunoglobulins where Ig-binding protein is present within the rough endoplasmic reticulum.

Simple assembly, e.g. haemoglobin

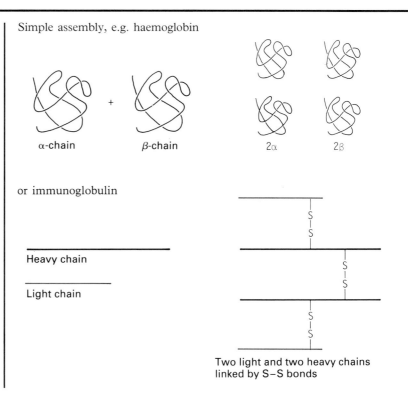

α-chain   +   β-chain   2α   2β

or immunoglobulin

Heavy chain

Light chain

Two light and two heavy chains linked by S–S bonds

A

Structure of proinsulin

Insulin is synthesized as a precursor proinsulin (e.g. pig proinsulin).

Insulin is closely related to relaxin and insulin-like growth factors-1 and -2 (somatomedins) which are synthesized in the liver and certain target tissues under the control of growth hormone (somatotrophin) secreted by the posterior pituitary (see also p. 40).

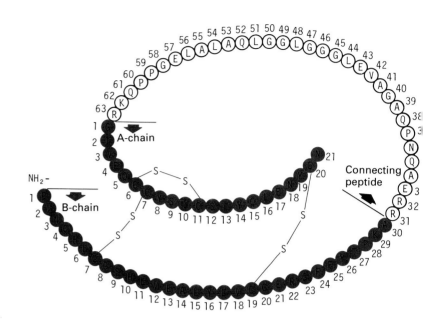

Removal of the *connecting peptide* (C-peptide) from proinsulin gives rise to insulin with the two chains A and B in the correct juxtaposition linked by S–S bonds. Proinsulin has only 5% of the biological activity of insulin and C-peptide is inactive. In the removal of the C-peptide the proteases PC2 and PC3 (see p. 40) are involved. These cleave on the carboxyl side of the basic amino acids (K and R). A carboxypeptidase B-like protease (carboxypeptidase H) then acts to remove the remaining RR. The products are, therefore, insulin and C-peptide without the terminal RR.

B

Various pro-proteins are known.

The role of the propeptide in serum albumin is not clear since there is a mutation which prevents its cleavage and hence proalbumin circulates in place of serum albumin. In spite of this, those affected appear to be normal.

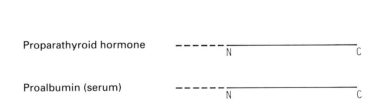

Two proproteins are shown. These together with many of the proteins in the blood coagulation cascade and numerous viral envelope proteins have their propeptide frequently linked through two adjacent basic amino acids. The propeptide is removed in or near the *trans* Golgi (see p. 253) by an enzyme called *furin* which is a member of a superfamily of PC proteases.

A

The template

Our knowledge of protein structure shows that the problem for the cell is to effect the linkage of amino acids by peptide bonds in the correct order.

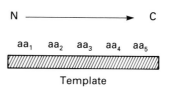

Template

From our knowledge of metabolism it might have been expected that polypeptides would be synthesized by a series of enzymic reactions using enzymes which were specific for each amino acid and peptide. It soon became apparent that too many enzymes would be involved and, moreover, there was no evidence for the presence in the cell of peptide intermediates. It was concluded that the amino acids must be assembled on a *template*, probably of nucleic acid. The early experiments showed that the synthesis of the polypeptide chain was in the direction $NH_2 \rightarrow COOH$.

2. STRUCTURE AND BIOSYNTHESIS OF NUCLEIC ACIDS

B

Replication, transcription and translation

The stages of nucleic acid and protein synthesis are usefully divided into four.

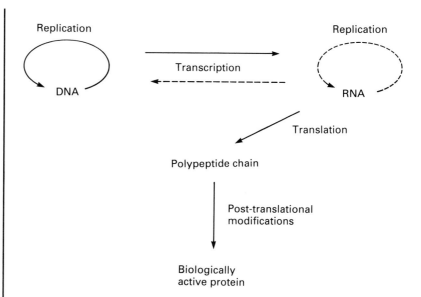

DNA is transcribed to RNA. One of the species of RNA is messenger RNA (mRNA), which is translated to give a polypeptide chain. Hence the template referred to in A above is mRNA. The other transcription products are transfer RNA (tRNA) and ribosomal RNA (rRNA). The dashed lines indicate activities that are essentially only concerned with RNA viruses. In some cases the viral RNA is merely replicated whereas in other cases the information contained within the base sequence of the viral RNA is used for the formation of DNA. Among the post-translational modifications are limited proteolysis, glycosylation, γ-carboxylation, phosphorylation and combination with prosthetic groups either by covalent or non-covalent bonds.

A

Nucleic acid and nucleotide structure

There are two kinds of nucleic acids:
deoxyribonucleic acid (DNA)
ribonucleic acid (RNA).
Nucleic acids are polymers, the monomeric unit being a *nucleotide*, so that all nucleic acids are *polynucleotides*. A nucleotide has three components:

phosphate — 5' sugar 1' — N-base

In DNA the sugar is deoxyribose, in RNA it is ribose. The nucleotides are linked in the polynucleotides through the formation of a phosphodiester bond. The basic structure of a polynucleotide is therefore:

Polyribonucleotide
('RNA')

base | base
—phosphate—ribose—phosphate—ribose—
(deoxyribose in 'DNA')

B

The nucleotides are linked by a *phosphodiester bond* between the 3' and 5' positions of each pentose.

The carbon atoms within the bases are numbered 1, 2, 3, etc., and those in deoxyribose and ribose 1', 2', 3', etc.)

Since the polynucleotide chain has polarity (i.e. is not the same in each direction), by convention the order of the nucleotides containing the different bases is read from the 5' end to the 3' end. Chains of opposite polarity may be said to be anti-parallel.

Illustrated is RNA.
In DNA the 2' OH is replaced by H.

C

There are typically only four different bases in either DNA or RNA. The only difference between DNA and RNA in this respect is that DNA contains *thymine* and RNA *uracil*.

As will be shown in tRNA, inosine is also present. This is the nucleoside of hypoxanthine, which is deaminated adenosine.

Pyrimidine bases

Cytosine (C)

Uracil (U)

(methyl-U = thymine (T))

Purine bases

Guanine (G)

Adenine (A)

Sites and directions for hydrogen bonding shown by ------→

A

Nucleic acids differ one from another by the sequence of the bases.

The characteristics of a particular nucleic acid are determined by the order in which the four different nucleotides occur in the polynucleotide chain. A polynucleotide is not symmetrical but has polarity, i.e. a 5' and 3' end. By convention the base sequence is specified 5' → 3'. Hence, pApCpGpApT specifies that the sequence in the polynucleotide starts with A at the 5' and has T at the 3' end. The p indicates the presence of a phosphate group and the capital letter the nucleotide.

B

DNA is usually *double-stranded*, the two chains being anti-parallel. The strands are coiled in the form of a *double helix* (see p. 87).

Double-stranded DNA with antiparallel strands

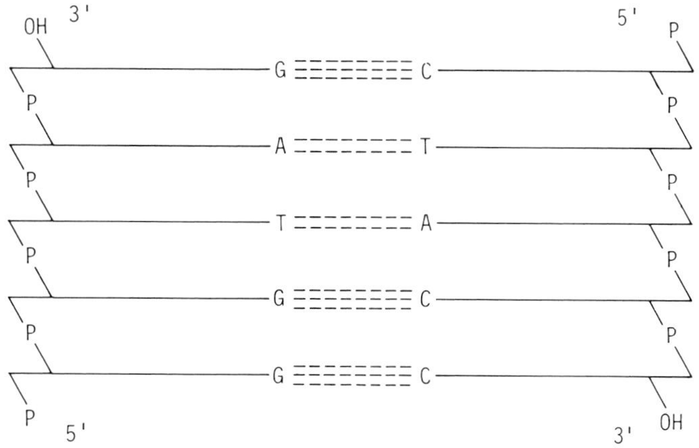

C

The proportion of the bases in DNA is such that A = T and G = C. This is in accord with the fact that A hydrogen bonds with T and G hydrogen bonds with C. This is known as 'base-pairing'.

Structure of base pairs in DNA

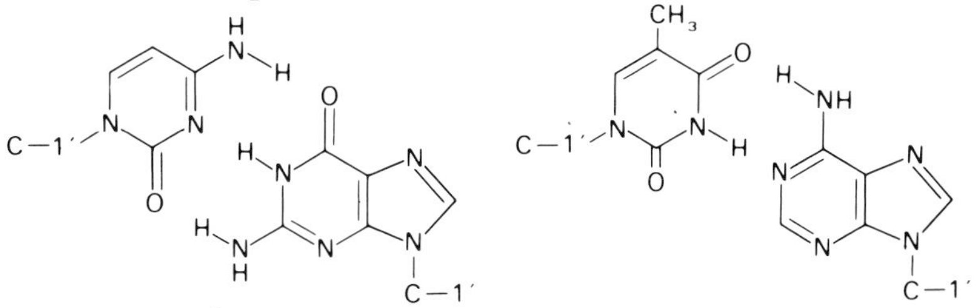

The base pair C:G is shown on the left, T:A on the right.

Note that the link between G:C is stronger than that between A:T. The antiparallel strands in DNA linked by base pairs are said to be *complementary*.

D

DNA comes in various sizes.

	No. of base pairs	Length
Viruses		
Polyoma (animal)	4 600	1.6 μm
Bacteriophage	53 000	16 μm
Bacteria		
Escherichia coli	3.4×10^6	1.2 mm
Human	1.2×10^{10}	200 cm

A *bacteriophage* is a virus which infects bacteria and is commonly known as a *phage*.

A

Nucleosides and nucleotides.

While a *nucleotide* has three components a *nucleoside* consists only of a sugar and a base:

$$5' \text{——} 1' \text{——} \text{N-base}$$
$$\text{sugar}$$

It follows that a nucleotide with three phosphate groups attached to a nucleoside at the 5' position of the sugar is named a nucleoside triphosphate, so that ATP stands for adenosine triphosphate:

$$\text{P—P—P—}5' \text{ ribose } 1'\text{-adenine}$$

Similarly, one obtains the names adenosine diphosphate (ADP) and adenosine monophosphate (AMP) and related terms for other bases.

B

Nomenclature in nucleic acids

Base	Nucleoside[*][†]	Nucleotide[†]
Purines		
Adenine	Adenosine (A)	Adenylic acid (AMP)
Guanine	Guanosine (G)	Guanylic acid (GMP)
Hypoxanthine	Inosine (I)	Inosinic acid (IMP)
Pyrimidines		
Cytosine	Cytidine (C)	Cytidylic acid (CMP)
Uracil	Uridine (U)	Uridylic acid (UMP)
(in RNA)		
Thymine	Thymidine (T)	Thymidylic acid (TMP)
(= 5-methyluracil in DNA		

[*]In polymers with repeating units the letters indicate nucleotides, e.g. poly(A) or poly(dT).
[†]When the sugar is deoxyribose the nucleoside or nucleotide is abbreviated dT, dAMP, etc. dNTP signifies unspecified deoxynucleoside triphosphate.

C

Hybridization of DNA

Double-stranded DNA can be denatured by heat and the strands can be re-annealed.

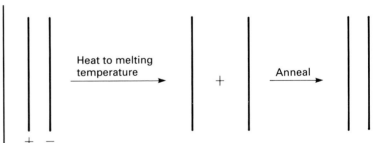

+ −

Native denatured to give single strands which   if complementary
DNA   hybridize to give
                                                 double-stranded
                                                 DNA when cooled

Complete denaturation of a molecule of DNA leads to separation of the two *complementary* strands. If a solution of *denatured* DNA is cooled quickly the denatured strands remain separated. If the temperature is held for some time just below T_m (the *melting temperature*), a process known as *annealing*, the native double-stranded structure can be reformed. This is termed *hybridization* when the nucleic acid is from different sources.

The DNA double helix

DNA exists as a *double helix* but may vary in conformation.

There is increasing speculation that DNA may change *in vivo* in conformation and that this may be important in gene expression. For example, transcription of DNA (see p. 97) probably occurs at the junction of B- and Z-DNA where unfolding takes place.

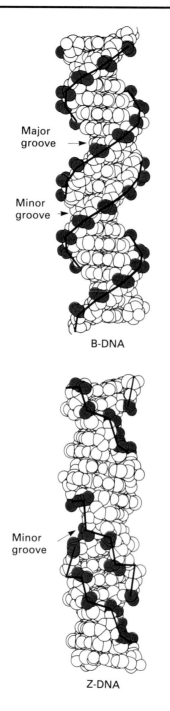

Major groove →

Minor groove →

B-DNA

Minor groove →

Z-DNA

DNA exists in solution as a right-handed helix known as the B-form. (The chains turn to the right as they move upward.) Note that this is characterized by a regular repeat of a minor and major groove.

The phosphate groups are shown in red.

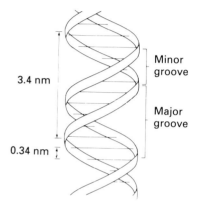

3.4 nm

0.34 nm

Minor groove

Major groove

Another quite different form of DNA has been found, Z-DNA, which has a left-handed helix. The irregularity of the Z-DNA backbone is illustrated by the heavy line that connects the phosphate groups.

A

The use of hybridization techniques

Hybridization can be used to compare the *homology* (degree of complementarity) of nucleic acids from different species. A DNA is broken into pieces of moderate length and is denatured. The denatured fragments from the two species are mixed and those that have similar nucleotide sequences tend to hybridize, whereas those where the sequences are very different do not. The reassociation of DNA fragments obeys second-order kinetics. A plot of the fraction of molecules reassociated versus log C_0t (where C_0 is the initial concentration of denatured DNA and t is time) is a convenient way of displaying data. The value of C_0t increases in direct proportion to the length of the DNA chain in the genome but it is very much decreased if the sequence of bases is highly repetitive (poly(U) and poly(A)). The slope at the midpoint gives an indication of the heterogeneity of the DNA fragments in solution.

B

Measurement of the rate of hybridization.

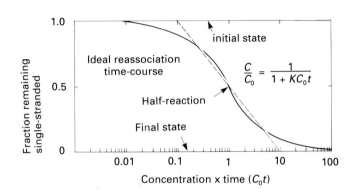

Time-course of an ideal, second-order reaction to illustrate the features of the log C_0t plot. The equation represents the fraction of DNA which remains single-stranded at any time after the initiation of the reaction. For this example, K is taken to be 1.0, and the fraction remaining single-stranded is plotted against the product of total concentration and time on a logarithmic scale.

C

An application: as the size of the DNA increases so the rate of reassociation is reduced.

Reassociation of double-stranded nucleic acids from various sources.

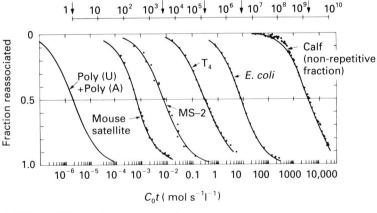

MS-2 and T$_4$ are from bacteriophages.

A

Biological activity and replication of DNA

The biological activity of DNA was first demonstrated by injecting mice with extracts of pneumococci.

R-strain, non-pathogenic; S-strain, pathogenic.

Two strains of pneumococci were used. The wild strain is pathogenic and is known as smooth or S because of the kind of colony it forms and the fact that it contains capsular polysaccharide. A mutant rough or R-strain lacks capsular polysaccharide and is non-pathogenic. The crucial experiment is illustrated, showing the effect of injection into the mice.

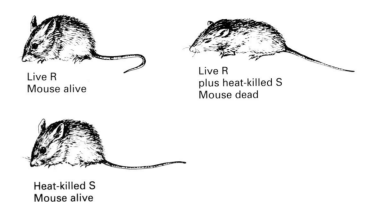

Live R
Mouse alive

Live R
plus heat-killed S
Mouse dead

Heat-killed S
Mouse alive

The heat-killed S was extracted and the active constituent shown by Avery and his colleagues in 1944 to be DNA. Since the blood of the dead mice contained live S-strain pneumococci, the DNA of the S-strain had *transformed* the live R-strain into live S-strain.

B

The infectivity of viral nucleic acid.

While the experiment illustrated represents the 'classical' way in which the biological activity of DNA was first demonstrated, with the passage of time simpler methods have been devised. In their most sophisticated form these are described under the section on recombinant DNA. Other methods involve the use of bacterial (bacteriophage) and plant viruses. The DNA from a bacteriophage can be shown to 'infect' a suitable host cell and cause it to 'lyse' as a result of phage replication. In some types of virus the genome is RNA rather than DNA. The RNA from a plant virus (e.g. tobacco mosaic virus) similarly can be shown to be infective. The primary structure of the coat protein of the resulting virus depends crucially on the structure of the infecting RNA.

C

Comparison of protein synthesis in prokaryotes and eukaryotes.

In general terms, nucleic acids and proteins are synthesized in all types of living cells in similar ways, but there are subtle differences as will be mentioned in the following pages. Such differences are particularly important in terms of the action of antibiotics. To be of clinical use these must be specific for either prokaryotes or eukaryotes (see p. 2). Hence, it is relatively easy to inhibit bacterial growth without toxicity to the human host. It is more difficult to treat a fungal (yeast) or protozoan (amoebic) infection since they, like their host, are eukaryotes.

Penicillin inhibits a key enzyme in *bacterial cell wall* synthesis and its action does not involve protein synthesis. Penicillin inhibits the cross-linking of different peptidoglycan strands.

DNA is replicated in the
cell by a process that
ensures that one of the
parent strands is present in
the daughter molecules.

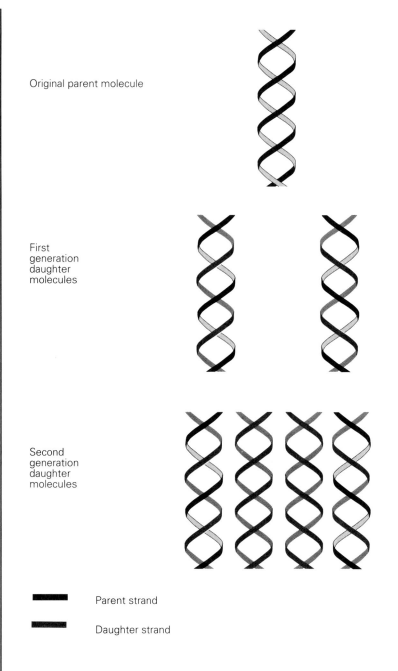

Original parent molecule

First
generation
daughter
molecules

Second
generation
daughter
molecules

Parent strand

Daughter strand

The experiments that proved that DNA was synthesized by a process of
semi-conservative replication in which one of the parental strands is
retained in the daughter molecules were done by Meselson and Stahl.
They grew *Escherichia coli* in the presence of the heavy isotope of N, ^{15}N.
They determined the density of the DNA and showed that after one
generation it was intermediate between the density of light DNA
(containing ^{14}N) and heavy DNA (containing ^{15}N).

A

The cell cycle

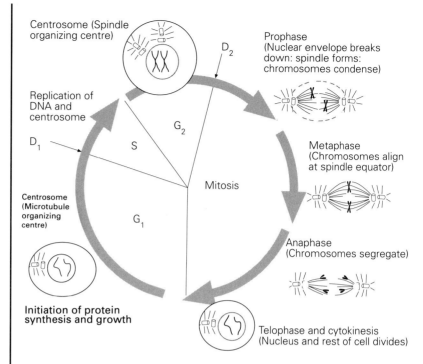

The cell cycle includes two phases: *interphase* during which the cell grows, and *mitosis* during which the nucleus and then the rest of the cell divides. Temporally, mitosis takes up about 40% in the cells of early embryos and less than 10% in most other cells. (Not as indicated in the figure.) The cycle often contains two decision-making points: D_1 (*start*) where the cell decides whether to replicate its DNA and D_2, whether to initiate mitosis. The protein Cdc2 is the main regulator of passage through both points. At D_1 a quiescent cell is said to have entered G_0.

B

Chromatin structure

In chromatin the histones are associated with the DNA in the form of nucleosomes.

A characteristic feature of the nucleus of eukaryotic cells, unlike prokaryotes, is the association of small basic proteins with the DNA. These are called *histones*. Four of the five different histones group in pairs to give a particle around which the DNA is wrapped. The product is known as a *nucleosome*. The nucleosomes are themselves coiled, so our concept of the structure of chromatin is as shown on the next page.

Types of histones

Type	Lys/Arg ratio	Number of amino acid residues	M_r ($\times 10^{-3}$)	
H1	20.0	215	21.0	
H2A	1.25	129	14.5	Components of nucleosomes
H2B	2.5	125	13.8	
H3	0.72	135	15.3	
H4	0.79	102	11.3	

A

A schematic view of the structure of *chromatin*.

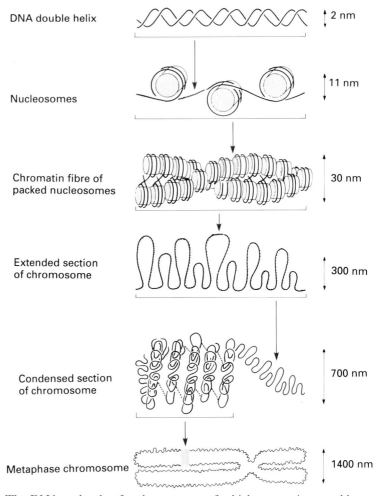

DNA double helix	2 nm
Nucleosomes	11 nm
Chromatin fibre of packed nucleosomes	30 nm
Extended section of chromosome	300 nm
Condensed section of chromosome	700 nm
Metaphase chromosome	1400 nm

The DNA molecule of a chromosome of a higher organism would stretch over many centimetres if laid out straight (see p. 85 D) and so must be highly folded to form the compact structure found in the nucleus. The DNA must also be organized into functional units which permit the reading of genes and their replication in a regulated manner.

B

The nucleosome is associated with five different types of histone.

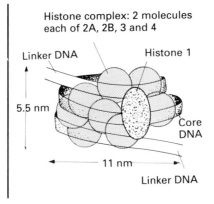

Histone complex: 2 molecules each of 2A, 2B, 3 and 4

Linker DNA

Histone 1

5.5 nm

Core DNA

11 nm

Linker DNA

DNA is organized in regular repeating units, nucleosomes, which contain successive segments of the double helix, about 200 base pairs long, associated with an octamer of four histone proteins, H3, H4, H2A and H2B. The DNA is wrapped in two turns around the octamer. At the point of entry and leaving, the DNA is sealed by another histone, H1. These repeating structural elements along the DNA form a filament which can coil into a helical or 'solenoidal' structure to form the second level of folding.

Chromosomes can be digested with a DNAase to produce a particle containing 14 helical turns of DNA and a histone protein core; this nucleosome core has been crystallized.

A

Replication and transcription of DNA and RNA

It is possible to isolate enzymes from cells which effect the synthesis of polynucleotides in the presence of deoxynucleoside triphosphates and a DNA template.

DNA is synthesized by enzymes called DNA polymerases, all of which utilize deoxynucleoside triphosphates as substrates, and the polynucleotide is formed in the direction 5' → 3'.

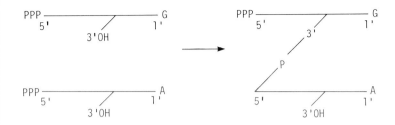

The horizontal line indicates the presence of deoxyribose and the numbers the identity of the carbon atoms of the sugar.

B

The role of parental DNA.

A DNA template is used to direct the order of bases in the newly synthesized polynucleotide such that it is *complementary* to the template DNA.

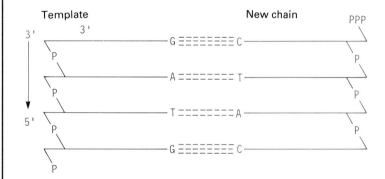

C

DNA ligase. An important enzyme for sealing nicks.

For the role of DNA ligase in DNA replication see page 95.

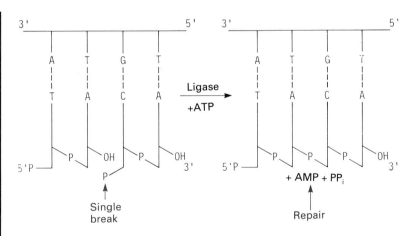

DNA *ligase* plays a role in the repair of a single break in a double-stranded DNA. Although ligase was discovered by studying DNA repair, it plays a key role in DNA synthesis in a normal cell.

In animal cells a covalent enzyme–AMP complex is formed as an intermediary. In bacteria, ATP is replaced by NAD^+.

A

In the living cell the most common mechanism of DNA replication is for it to proceed in both directions along the two strands of DNA simultaneously. In the simplest case of *Escherichia coli*, unwinding and replication of the double-stranded circular DNA leads to the production of a structure that resembles a circle with an inner loop. In this case, replication is from a single initiation point.

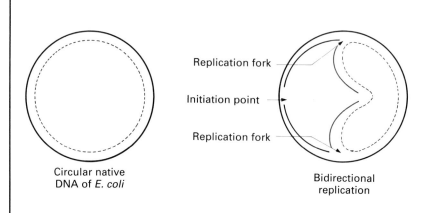

Circular native DNA of E. coli

Replication fork

Initiation point

Replication fork

Bidirectional replication

The problem posed is that all known DNA polymerases cause the synthesis of new polynucleotides in the direction 5' → 3'. None have been found which synthesize strands in the opposite direction.

B

In order to explain the apparent growth of both strands of DNA in the same direction, it is postulated that at least one of the strands is first synthesized in fragments.

This process is explained in greater detail on page 95 B.

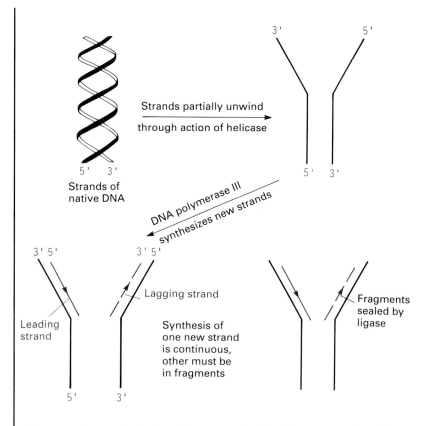

Strands of native DNA

Strands partially unwind through action of helicase

DNA polymerase III synthesizes new strands

Leading strand

Lagging strand

Synthesis of one new strand is continuous, other must be in fragments

Fragments sealed by ligase

The experiments that led to this concept involved the examination of DNA shortly after its synthesis when a number of small fragments were found. The work was done by Okazaki and the fragments are therefore named after him.

A

The role of RNA in DNA biosynthesis.

It came as a surprise that DNA synthesis also required the cell to be synthesizing RNA. A specific enzyme, 'primase', is responsible.

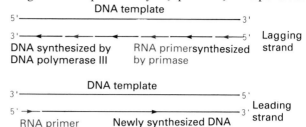

On the *lagging strand* the RNA primer is removed by DNA polymerase I, the gaps filled by the same enzyme and all fragments sealed by ligase. On the *leading strand* the RNA primer is synthesized by RNA primase and removed by DNA polymerase I, the space filled by DNA polymerase I and the two DNA fragments joined by ligase. Both DNA strands appear to grow in the same direction, i.e. from left to right.

B

DNA biosynthesis is a multienzyme process. Scheme for enzymes operating at one of the forks in the bidirectional replication of an *Escherichia coli* chromosome.

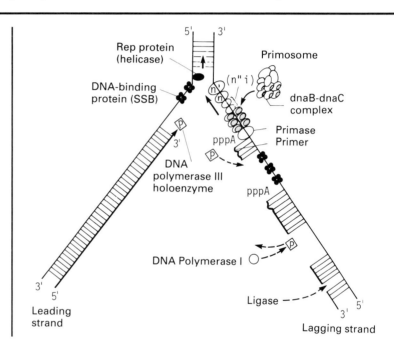

Three DNA polymerases have been isolated from *E. coli*. Polymerase III is for synthesis, polymerase I is mainly for editing to ensure faithful replication but also has exonuclease activity and polymerase II is membrane bound. At least 15 replication proteins (comprising over 30 polypeptides) have been characterized. The dnaA protein binds to DNA to be joined by a complex of the dnaB and dnaC proteins. dnaB is a helicase which catalyses the ATP-driven unwinding of the double helix. The unwound portion of DNA is then stabilized by single-stranded binding protein (SSB). An RNA primer is formed by primase in a multisubunit assembly called the primosome. At the end of replication this will be removed by DNA polymerase I. New DNA is synthesized by DNA polymerase III holoenzyme consisting of eight polypeptide chains. The DNA ahead of the polymerase on the leading strand is unwound by helicase (the Rep protein). On the lagging strand the gaps between the fragments of DNA are filled by DNA polymerase I and the fragments joined by DNA ligase.

In eukaryotes at least five DNA polymerases have been identified. These are designated by Greek letters. α and δ are involved in the replication of nuclear DNA, β with editing and repair, and γ with the replication of mitochondrial DNA.

The metabolism of nucleotides

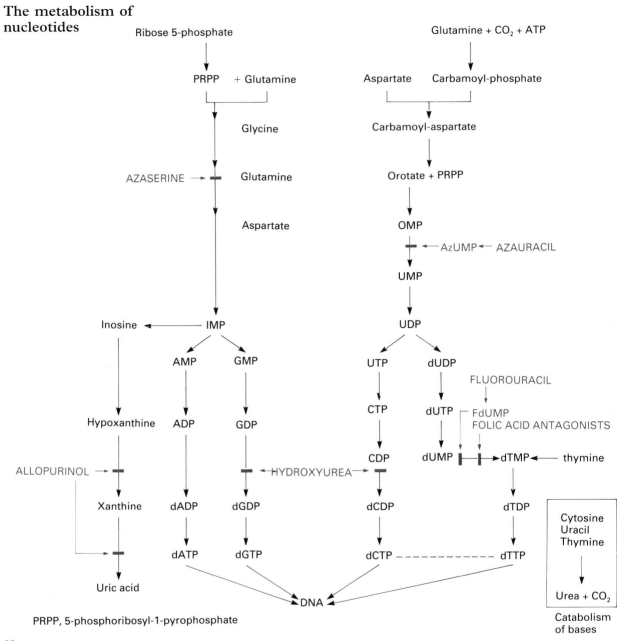

PRPP, 5-phosphoribosyl-1-pyrophosphate

Catabolism of bases

Note.
1. In the conversion of dUMP to dTMP the methyl donor is N^5,N^{10}-methylenetetrahydrofolate. The regeneration of tetrahydrofolate is accomplished by dihydrofolate reductase, which is inhibited by *aminopterin* and *methotrexate*. For tetrahydrofolates see page 137.
2. Free purine bases formed by the degradation of nucleic acids and nucleotides are converted to purine nucleotides by a *salvage pathway*. The ribose phosphate moiety of PRPP is transferred to the purine. Thus

$$\text{Adenine} + \text{PRPP} \rightarrow \text{Adenylate} + \text{PP}_i$$
$$\text{Hypoxanthine} + \text{PRPP} \rightarrow \text{Inosinate} + \text{PP}_i$$
$$\text{Guanine} + \text{PRPP} \rightarrow \text{Guanylate} + \text{PP}_i$$

Gout, which is characterized by elevated levels of uric acid in the blood, can be alleviated by the drug allopurinol, the action of which is shown. This results in a reduction in the formation of uric acid and an increase in the excretion of the more soluble xanthine and hypoxanthine.

A

Transcription

Transcription is said to be *asymmetric* in that only one strand of DNA is *transcribed*, and *conservative* in that the double-stranded DNA template is *conserved*.

Three kinds of RNA are produced:

  Ribosomal RNA (rRNA)
  Transfer RNA (tRNA)
  Messenger RNA (mRNA).

The base sequence of the synthesized mRNA is complementary to that of the DNA strand used as a template.

In *Escherichia coli* the relative amount of the three types of RNA are rRNA 80%, tRNA 15%, mRNA 5%.

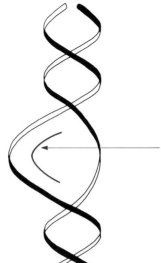

Double-stranded DNA template
Clearly there must be a precise mechanism to allow for the initiation of transcription at a certain point on one strand of the template DNA.

Newly synthesized RNA formed by RNA polymerase

The DNA strands separate a little at the appropriate point.

The antibiotic actinomycin D inhibits transcription by binding to the DNA template and rifampicin by binding to the DNA-dependent RNA polymerase.

B

Viral replication

The RNA of an RNA virus is replicated by RNA-dependent RNA polymerase.

A viral RNA may be either a (+) strand, i.e. an mRNA as in polio, or a (−) strand as in measles, rabies and influenza.

An RNA virus such as poliomyelitis must arrange for the replication of many (+) strands. It does this by forming a (−) strand from which many (+) strands are formed by asymmetric synthesis.

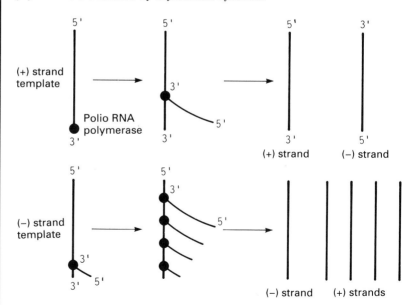

The polymerase is a multisubunit protein (4) in which only one of the protein subunits is virus-specific. The other subunits are provided by the host cell. RNA virus other than polio is replicated by a variety of different methods.

A

Reverse transcriptase

RNA oncogenic virus contains an RNA-directed DNA polymerase. The viral RNA is (−) strand.

The formation of double-stranded DNA complementary to viral RNA by *reverse transcriptase*.

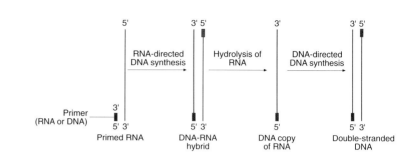

B

The copy of the viral RNA is integrated into the host chromosome, becoming a *provirus*.

Such a virus is known as a *retrovirus*.

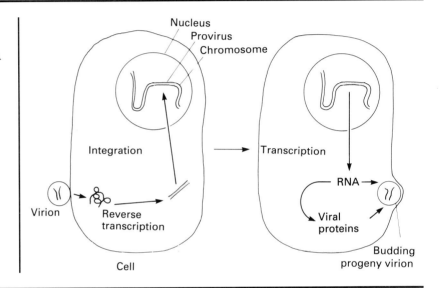

C

Structure of HIV-1

The structure of human immunodeficiency virus (HIV-1), the causative agent for acquired immunodeficiency syndrome (AIDS), a retrovirus.

For the role of retroviruses as oncogenes, see the next page.

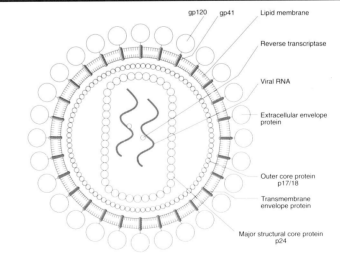

The figure shows the structure of HIV-1. The HIV virion is about 100 nm in diameter. The distribution of the proteins is shown. HIV enters T4 lymphocytes by the interaction of gp120 with a plasma membrane receptor CD4. The two membranes fuse and the viral core is released into the cytosol, which results in the destruction of T4 lymphocytes, a vital contributor to the immune response. The primary structure of gp120 undergoes rapid mutation, which makes a preparation of an effective vaccine difficult.

A

Retroviruses as oncogenes

Retroviruses are the only cancer-producing RNA viruses. (DNA viruses such as polyoma virus, simian virus 40 (SV40) and Epstein–Barr virus (EBV) also induce tumours.) Retroviruses that cause tumours are called *oncogenic RNA viruses (ONCORNA)*. They act either by introducing an *oncogene* into the host DNA, which leads to the synthesis of altered proteins, or they cause the production of excessive quantities of normal proteins that play a key role in growth regulation. The normal cells are said to be *transformed* to *neoplastic cells* that not only grow faster than normal but are also immortal. Some ways in which oncogenes may act are described on page 273. Normal cells contain genes which may be mutated (often by only one base change) to become cancer-producing, such genes are called proto-oncogenes, which unlike oncogenes contain introns (see p. 121). It is possible that viral oncogenes were derived from proto-oncogenes.

B

Action of antibiotics and antimetabolites

The site of action of various antibiotics and other substances that inhibit DNA polymerase, RNA polymerase and the synthesis of nucleotides. For further explanation, see pages 96 and 108.

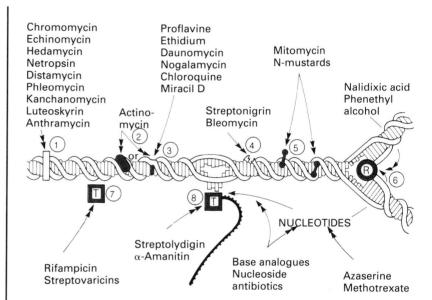

Chromomycin
Echinomycin
Hedamycin
Netropsin
Distamycin
Phleomycin
Kanchanomycin
Luteoskyrin
Anthramycin

Proflavine
Ethidium
Daunomycin
Nogalamycin
Chloroquine
Miracil D

Mitomycin
N-mustards

Nalidixic acid
Phenethyl
alcohol

Actino-
mycin

Streptonigrin
Bleomycin

Rifampicin
Streptovaricins

Streptolydigin
α-Amanitin

Base analogues
Nucleoside
antibiotics

Azaserine
Methotrexate

NUCLEOTIDES

1. Non-covalent binding to DNA by unknown mechanism
2. Intercalation without distortion of helix structure
3. Intercalation between bases with distortion of helix structure
4. Breaking of strands
5. Alkylation and cross linking

6. Inhibition of DNA polymerase (replication)
7. Inhibition of initiation of RNA polymerase (transcription), not effective if added after initiation
8. Inhibition of RNA polymerase at any time.

T, transcription; R, replication.

3. BIOSYNTHESIS OF PROTEINS

A

Amino acid adenylates, tRNA

The first step in protein synthesis involves the activation of the amino acids.

Since polypeptides are polymers, energy must be utilized in their formation. Hence an 'activated' amino acid was sought. There is a specific enzyme for the activation of each of the 20 different amino acids.

The 'activation' of an amino acid by reaction with ATP yields an amino acid adenylate and pyrophosphate.

B

The second step involves the transfer of the activated amino acid to *tRNA*.

The same enzyme is involved as in step 1. There is at least one specific tRNA for each of the 20 amino acids but all have CCA at the 3' end.

The amino acid residue is attached to tRNA by an ester bond which facilitates its subsequent utilization in the formation of a peptide bond.

A

Properties of ribosomes

Ribosomes may be characterized by their sedimentation constant.

The site of protein synthesis in the cell is the cytoplasmic ribosome. This contains Mg^{2+} ions, proteins and RNA. Ribosomes consist of two subunits one about twice the size of the other.

Sedimentation constants of ribosomes and their subunits

	Intact ribosome	Large subunit	Small subunit
Prokaryotes	70S	50S	30S
Eukaryotes	80S	60S	40S

The ribosomes present in mitochondria are smaller than those in the cytoplasm. They vary in size according to source but are more like 70S than 80S ribosomes.

B

Characteristics of rRNA.

Subunit	S of RNA	M_r	Subunit	S of RNA	M_r
50S	23S	0.98×10^6	60S	28S	1.7×10^6
	5S	40 000		5.85	51 000
30S	16S	0.9×10^6		5S	39 000
			40S	18S	0.7×10^6

C

The genetic code

Codons which code for an amino acid are called *sense* codons. The three codons that do not code for an amino acid are called *nonsense* codons.

Second base of codon:

	U	C	A	G	
U	UUU] Phe UUC] UUA] Leu UUG]	UCU ⌐ UCC UCA Ser UCG ⌐	UAU] Tyr UAC] UAA] Term* UAG]	UGU] Cys UGC] UGA Term* UGG Trp	U C A G
C	CUU ⌐ CUC CUA Leu CUG ⌐	CUU ⌐ CCC CCA Pro CCG ⌐	CAU] His CAC] CAA] Gln CAG]	CGU ⌐ CGC CGA Arg CAG ⌐	U C A G
A	AUU ⌐ AUC AUA Ile AUG Met+Init.†	ACU ⌐ AAC ACA Thr ACG ⌐	AAU] Asn AAC] AAA] Lys AAG]	AGU] Ser AGC] AGA] Arg AGG]	U C A G
G	GUU ⌐ GUC Val GUA ⌐ GUG Val+Init.†	GCU ⌐ GCC GCA Ala GCG ⌐	GAU] Asp GAC] GAA] Glu GAG]	GGU ⌐ GGC GGA Gly GGG ⌐	U C A G

First base of codon. Third base of codon.

Theoretical consideration led to the prediction that the code, for the translation of the base sequence in mRNA into the order of amino acids in a polypeptide, would be a 'triplet' of bases called a codon. The code shows that 61 triplets code for an amino acid. Three triplets (marked *) indicate the termination of a polypeptide chain, two other triplets are used for initiation as well as for Met and Val, respectively (marked †). The genetic code has generally been considered to be universal in that it appears to be valid for all living cells whether eukaryotic or prokaryotic. (GUG seems seldom to be used for initiation.) The genetic code for mitochondria differs from that indicated above. Thus, in humans, UGA is used for Trp rather than Term., AUA for Met rather than Ile and AGA for Term. rather than Arg.

A

The reality of the code *in vivo*.

We may check the 'code' in two ways. First to see whether it accounts for single amino acid changes in a haemoglobin chain, such as occurs in sickle-cell disease, on the basis of a *single-point mutation*. It does. Secondly, we can now determine the primary structure of mRNA and relate it to the structure of a protein as shown in a hypothetical peptide. The polypeptide and mRNA are found to be *collinear*. The mRNA is read 5' → 3'

B

The gene for *haemoglobin S* is derived from that for normal adult haemoglobin A by a single-point mutation.

HbS differs from HbA by the change of glutamic acid at position 6 in the β-chain of HbA for valine in HbS.

	HbA Glu	HbS Val
Possible } codons }	G*A*A G*A*G	G*U*A G*U*G

In each case A has been changed to U, i.e. only one base has been changed. In fact the base change is GAA to GUA.

C

Collinearity of polypeptide and mRNA.

Recognition of the initiator codon determines the *reading frame*, i.e. which bases will form the triplets for successive codons.

NH₂ term   COOH term

Met Pro Ser Ala Gly Gly
AUGCCGUCGGCGGGCCGUUAG
5'   3'

The sequence of bases starts with AUG, which is not only the codon for Met but also is used as the initiator codon (see later), and ends with UAG, which is one of the three codons used for termination. (No amino acid is specified as it is a *nonsense* codon.) In the mRNA for any particular protein it is found that several codons for a particular amino acid such as Gly may be used as indicated. Thus, there is *degeneracy* within a particular mRNA.

D

Structure of eukaryotic mRNA

A typical mRNA contains a region at the 5' end which is not translated and is hence designated as the 5' non-coding region. (In bacteria there is evidence that the base sequence is complementary to the 3' end of the RNA of the smaller of the two ribosomal subunits (16S RNA) and hence it is assumed that the mRNA non-coding region has a role in the association of the mRNA with the ribosomal subunit.) There is also a non-coding region at the 3' end of the mRNA. As described on page 121 this end usually bears a poly(A) tail. Many eukaryote mRNAs also have a 'Cap' of 7-methylguanosine at the 5' terminus. This is attached by the triphosphate bridge to the 5'-terminal nucleotide.

5' Cap   Non-coding 5'   Coding   Non-coding 3'   Poly(A)   3'

One or more bases in mRNA may be modified in vivo, a process known as *RNA editing*. Most of the known cases concern mitochondria but in the mRNA for apolipoprotein B (see p. 222B) a cytidine is deaminated to a uridine. Thus, the codon for glutamine (CAA) becomes a stop codon (UAA), and a shortened form of the protein is the result of translation. The site-specific cytidine deaminase involved has no energy or cofactor requirements.

A

The role of tRNA

In the early studies there was much speculation as to the way any particular amino acid would become associated with its codon on the mRNA. Crick predicted that it would be through the medium of a small RNA which would bear an *anticodon*. This hypothetical RNA was called *adaptor RNA*. tRNA fulfils the role of the adaptor RNA. Thus, it was shown that tRNA bearing an amino acid has the unique property of locating itself at the correct codon. tRNA must therefore bear an anticodon. The structure of the anticodon XYZ (3' → 5') can be predicted from our knowledge of base pairing (A-U, G-C).

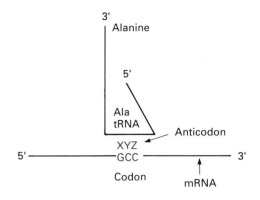

B

The structure of alanine tRNA.

A common feature of all tRNAs is that they contain many intramolecular base pairs and unusually modified bases, e.g. methylguanosine (MeG).

From the genetic code one might have expected there to exist 61 different tRNAs (one for each sense codon) but there are probably only about 40 different tRNAs since one anticodon may pair with more than one codon. The so-called 'Wobble' hypothesis predicted GC, AU pairing at the first two codon bases but imprecise pairing at the third codon base.

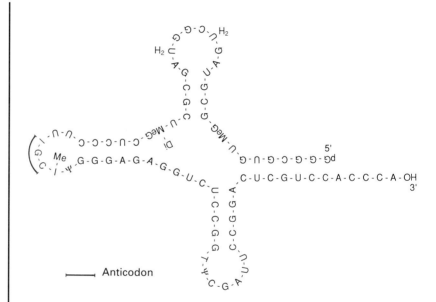

Having determined the primary structure of tRNA, Holley studied the way in which the maximum number of intramolecular base pairs could be formed. All the bases except those shown in the loops could be so base-paired and the predicted structure was a clover leaf. This structure has been confirmed from the X-ray crystallography of some tRNA molecules. As predicted, one finds a triplet of bases which could serve as an anticodon in a loop (the anticodon loop). I stands for inosine which is deaminated to adenosine (A). Inosine can pair with U, C or A. For alanine we have:

5' Codons	3'	GCU	GCC	GCA	GCG
3' Anticodons	5'	CGI	CGI	CGI	CGC

Hence two different tRNAs suffice for alanine.

A

The ribosome cycle—polyribosomes

Several ribosomes are attached to a molecule of mRNA to form a *polyribosome* (polysome for short). During the formation of a polypeptide, ribosomes move with respect to the mRNA and leave it at the 3' end on completion of the polypeptide chain. Ribosomal subunits (see (p.101) rejoin the mRNA at the 5' end. There are three steps in the synthesis of a polypeptide. *Initiation, elongation* and *termination.*

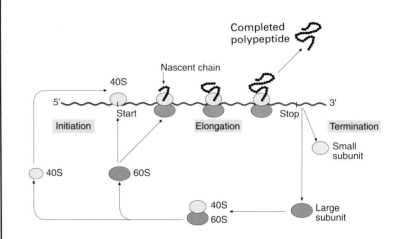

B

The ribosome cycle—initiation in eukaryotes

The initiation site on the mRNA is at a specific AUG codon. *Initiation factors* (eIFs) associate with the mRNA and are joined by the 40S subunit together with Met-tRNA and guanosine triphosphate (GTP) to give a complex with the AUG correctly aligned. This is then joined by the 60S subunit, with the release of guanosine diphosphate (GDP), to give the 80S initiation complex, The process in prokaryotes is essentially similar with fewer different IFs.

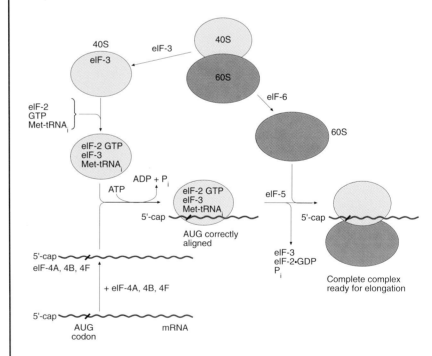

A

The ribosome in chain elongation

Chain elongation involves a shuttle movement of tRNA between sites P and A.

When the two tRNAs are assembled on the ribosome a peptide bond is formed by a nucleophilic attack of the α-amino group attached to the incoming tRNA on the carboxyl group of the tRNA to which the nascent chain is attached. The tRNA is then released. The aminoacyl tRNA normally enters site A.

An elongation factor (such as EF-Tu in bacteria) positions the aminoacyl tRNA in the A site.

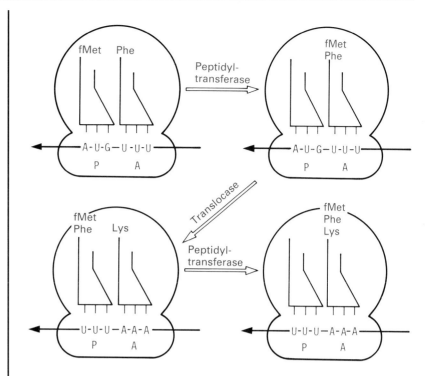

The enzyme peptidyltransferase catalyses the transfer of the incoming amino acid to the COOH end of the nascent peptide chain. Translocase is the enzyme that catalyses the movement of the new tRNA with its nascent peptide from site A (aminoacyl) to site P (peptidyl), known as translocation.

B

The ribosome in chain initiation

The problem of chain initiation.

The problem of chain initiation is to get the tRNA bearing the amino acid which will serve as the N-terminal amino acid of the nascent peptide into site P. In eukaryotes, Met, and in prokaryotes, formyl Met, always function as the N-terminal amino acid. There are two different tRNA molecules for Met even though there is only one codon. The tRNA which has the unique property of entering site P is called initiator tRNA. A Met bound to this tRNA may be formylated in the presence of an enzyme, as it is in prokaryotes, but in any case is designated tRNA^{fMet}. Hence, in principle the mechanism of chain initiation in prokaryotes and eukaryotes is similar, the eukaryotes lacking only the formylating enzyme. (The mechanism in mitochondria is more like that in prokaryotes as explained on page 115.) After completion of the chain in prokaryotes the formyl group is always removed and the N-terminal methionine also if it is not the terminal amino acid residue of the mature protein. In eukaryotes the methionine is removed in appropriate cases when about 30 amino acid residues have been added to the chain.

A

The ribosome in chain termination

Release factors are involved.

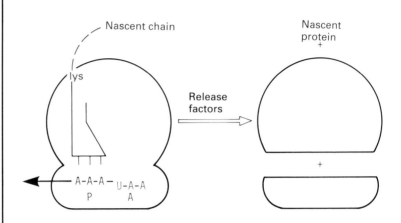

When UAA or one of the other nonsense termination codons (UAG, UGA) occupies site A, a protein factor (RF) binds to site A in the presence of GTP and catalyses the termination reaction. The reaction involves the hydrolysis of the peptidyl tRNA ester bond, the hydrolysis of GTP and the release of the completed polypeptide chain, the deacylated tRNA and the ribosome from mRNA. In prokaryotes there are three release factors, but only one is required in eukaryotes. Following the release of the polypeptide chain the ribosome dissociates into its two subunits, ready for another round of protein synthesis.

B

The site of action of antibiotics

Puromycin interrupts protein synthesis by virtue of the similarity of its structure to the 3' end of amino acyl tRNA.

Puromycin enters site A on the ribosome and a C-terminal puromycinyl polypeptide is released from the ribosome.

Many antibiotics are effective because they inhibit protein synthesis and are specific for either eukaryotes or prokaryotes. *Puromycin* has been particularly important in research on the mechanism of protein synthesis but is not specific with respect to cell type.

A

Various antibiotics inhibit protein synthesis by interfering with either peptidyltransferase or translocase.

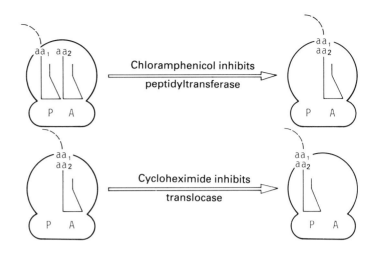

Chloramphenicol is specific for prokaryotes and mitochondrial protein synthesis. This accounts for the toxic effect of chloramphenicol in causing aplastic anaemia.

Cycloheximide is specific for eukaryotes.

B

The site of action of antibiotics.

G-factor and T-factor are proteins involved in elongation.

Resistance to a wide variety of chemotherapeutic drugs is commonly associated with high expression levels of a single protein, the 170 kDa P-glycoprotein encoded in humans by the *MDR1* gene. The protein pumps drugs out of cells by an ATP-dependent process.

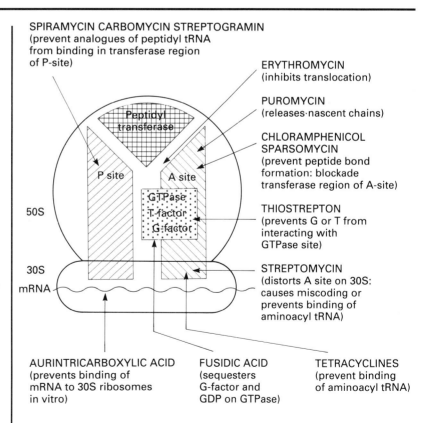

The action of various drugs

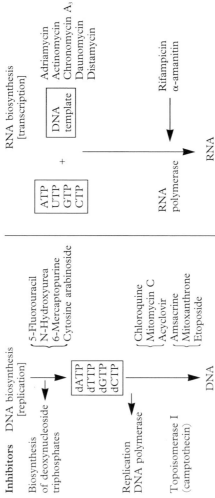

Acyclovir (acycloguanosine)

AZT (zidovudine) (Retrovir) (3'-azido-3'-deoxythymidine)

Inhibitors   DNA biosynthesis [replication]

Biosynthesis of deoxynucleoside triphosphates
{ 5-Fluorouracil, N-Hydroxyurea, 6-Mercaptopurine, Cytosine arabinoside }

dATP dTTP dGTP dCTP → DNA

Replication DNA polymerase
{ Chloroquine, Mitomycin C, Acyclovir }

Topoisomerase I (camptothecin)
{ Amsacrine, Mitoxanthrone, Etoposide }

RNA biosynthesis [transcription]

ATP UTP GTP CTP + DNA template
{ Adriamycin, Actinomycin, Chronomycin A₃, Daunomycin, Distamycin }

RNA polymerase → RNA
{ Rifampicin, α-amanitin }

Ribosomal processes (transcription)

Initiation → Recognition of codon → Peptidyl-transferase → Translocase → Termination

{ Tetracycline, Kasugamycin }
{ Streptomycin, Gentamicin, Neomycin, Kanamycin }
{ Lincomycin, Chloramphenicol, Erythromycin, Puromycin }
{ Fusidic acid, Erythromycin, Cycloheximide, Emetine }

1. Action at the bacterial ribosome (prokaryotic cell ribosome):
   a. Streptomycin, neomycin, kanamycin and kasugamycin act at the 30S subunit
   b. Chloramphenicol, erythromycin and puromycin act at the 50S subunit.
2. Action at the eukaryotic cell ribosome by cycloheximide and emetine.

Distinction is made between five types of inhibitors of DNA biosynthesis according to site of attack:
1. Inhibitors which reversibly inhibit the DNA polymerase reaction. Aromatic compounds which form a complex with the DNA double helix and thus inhibit replication, e.g. chloroquine. Nalidixic acid also inhibits but the site of action is uncertain.
2. Inhibitors which react with the DNA with formation of covalent bonds enter into so-called cross-links with the two DNA strands necessary for semi-conservative replication, e.g. the alkylating agents including nitrogen mustard, chlorambucil, cyclophosphamide; and myleran, mitomycin C. The platinum drugs, cisplatin and carboplatin, induce both inter- and intrastrand cross-links. Hexamethylmelamine and dacarbazine (DTIC) methylate DNA.
3. Acyclovir is a guanine derivative which is phosphorylated in herpes virus infected cells much faster than in uninfected cells, due to the action of a herpes-specific thymidine kinase. The product, acycloguanosine triphosphate, inhibits herpes virus DNA polymerase more effectively than cellular DNA polymerase.
4. Inhibitors such as N-hydroxyurea, 5-fluorouracil, aminopterin and methotrexate inhibit a certain step in the biosynthesis of deoxyribonucleoside triphosphates, and hence indirectly DNA synthesis, by lowering the pool concentration of deoxynucleotides (see p. 96).
5. AZT undergoes phosphorylation in human T-cells to a nucleoside 5-triphosphate which competes with TTP and serves as a chain-terminating inhibitor of HIV reverse transcriptase.

4. CONTROL OF PROTEIN SYNTHESIS

The operon

The scheme for the control of the *lac* operon in bacteria has formed the basis for our understanding of the mechanism of action of hormones on protein synthesis.

Translation can also be prevented by the use of a plasmid (see p. 118) which results in the synthesis of a mRNA that is complementary to the mRNA the translation of which it is desired to inhibit. Such an RNA is described as an *antisense mRNA*. Isolated cells can also be treated with single-stranded DNA and chemically modified DNA which is complementary to mRNA.

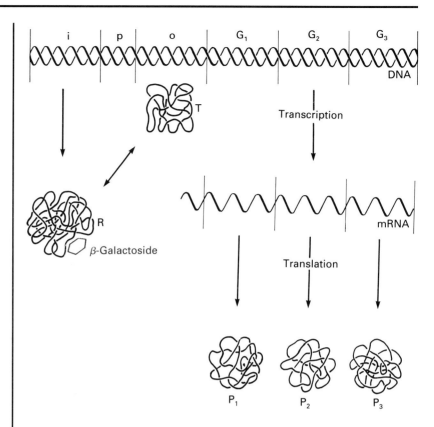

The Jacob–Monod scheme for the induction of enzymes in bacteria. R, repressor protein in state of association with the galactoside inducer shown by the hexagon; T, repressor protein in state of association with operator segment (o) of DNA; i, 'regulator gene' governing synthesis of the repressor; p, 'promoter' segment, point of initiation for synthesis of mRNA; G_1, G_2, G_3, 'structural' genes governing synthesis of the three proteins in the system marked P_1, P_2, P_3—one of these proteins is β-galactosidase, the others are permease and transacetylase.

Whilst the repressor (protein) is bound to DNA, transcription cannot occur, since the site of initiation of transcription is at the promoter segment. The action of the β-galactoside is to cause the repressor to leave the DNA so that transcription and subsequently translation can occur. Since the normal state is one in which the *lac* operon is inhibited the phenomenon as described here is known as *negative control*.

The *lac* operon is, however, also subject to *positive control*. Thus, cAMP binds to a *catabolite gene activation protein* (CAP) and the complex stimulates transcription by enhancing the tightness of binding of RNA polymerase to the promoter. The circumstance whereby glucose is the preferred energy source, and in its presence there are only very low levels of β-galactosidase and other catabolic enzymes, is explained by the phenomenon of *catabolite repression*. The presence of glucose lowers the concentration of cAMP and hence there is an absence of the *positive control* referred to above.

A

Promoters and enhancers of transcription in eukaryotes

The organization of the genome

The TATA box is sometimes known as the Hogness box.

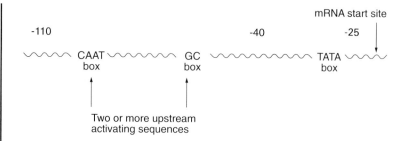

General features of promoters for mRNA precursors in eukaryotic cells are illustrated above.

The first nucleotide (the *start site*) of a DNA sequence is denoted as $+1$, followed by $+2$, etc. The nucleotide preceding the start site is denoted -1. These refer to the coding (*sense*) strand, not the template (*antisense*) strand. The coding strand has the same sequence as the mRNA except for T in place of U. Promoters for RNA polymerase are on the 5' side of start. As shown there are several of these, the *TATA* and *CAAT* boxes are very common, as is the *GC* box. Such boxes are necessary for promoter activity but not sufficient. *Additional elements* are located between -40 and -110. For the polymerase to recognize the promoter, *transcription factors* are also needed. The activities of many promoters are increased by *enhancers* which may bind to DNA over several thousand base pairs. The binding may be on the 5' side (upstream) or on the 3' side (downstream) or even within a gene. Such enhancers are often tissue-specific.

B

Protein–DNA recognition

Much effort is being devoted to discovering how transcription factors recognize regulatory sequences in genes. The first such motif to be discovered was that present in the proteins which bind to homeotic genes (these control the architectural plan of the embryo). This motif is known as the 'helix–turn–helix', as it contains a short region which conforms to an α-helix followed by a β-turn and then another α-helix. This motif is present not only in the homeobox of *Drosophila* but also in bacterial regulatory proteins such as Cro.

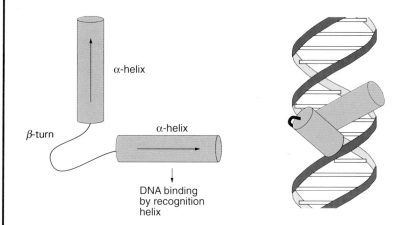

In its binding to DNA one helix lies across the major groove while the second helix lies partly within the major groove where it can make specific contacts with DNA bases.

Factors that are the product of genes other than that which they control, which is the usual case, are said to be *trans*-acting. Products of the same gene which is controlled are said to be *cis*-acting, which is unusual.

A

Protein–DNA recognition—
cont.

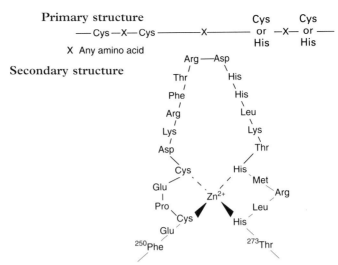

Another motif is known as the '*zinc finger*'. The structure shown contains a pair of Cys residues followed by about 12 amino acid residues followed by another pair of Cys residues. The Cys pairs coordinate with Zn^{2+} so that the other amino acids form a finger-like protrusion. This structure is repeated nine times in the transcription factor TFIIIA which binds to the gene for 5S RNA. In the control of gene expression by steroid hormones the specific receptor protein contains a hormone-binding region and a DNA-binding section. The latter domain resembles two conventional zinc fingers except that one His pair is replaced by another Cys pair and the finger is a little larger.

B

Model of the binding of the zinc fingers in TFIIIA to the 5S DNA.

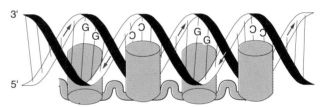

Note that adjacent fingers make contact with the DNA from opposite sides of the helix.

C

Comparison of the structures of compounds that are thought to act by binding to the hormone-binding region of the specific receptors described above.

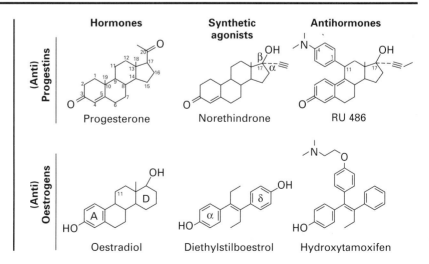

Tamoxifen itself is an anti-oestrogen that is much used in the prevention and treatment of breast cancer. RU 486 (mifepristone) is used in combination with prostaglandin analogues for the termination of pregnancy.

5. MOLECULAR CELL BIOLOGY

A

Secretory proteins

Secretory proteins are synthesized on the rough-surfaced endoplasmic reticulum from which they move through the cell. They may be modified either by the attachment of prosthetic groups or by partial proteolytic cleavage (examples of *post-translational modification*).

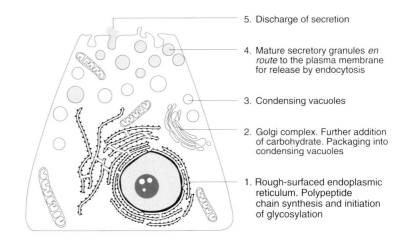

5. Discharge of secretion

4. Mature secretory granules *en route* to the plasma membrane for release by endocytosis

3. Condensing vacuoles

2. Golgi complex. Further addition of carbohydrate. Packaging into condensing vacuoles

1. Rough-surfaced endoplasmic reticulum. Polypeptide chain synthesis and initiation of glycosylation

The binding of polyribosomes to the endoplasmic reticulum is an essential step in the synthesis of proteins destined for secretion.

B

Polyribosomes for secretory and housekeeping proteins.

The structure of many 'signal' peptides is now known and they all contain a run of hydrophobic amino acids residues.

In order to determine the nature of the primary translation product of the mRNA of a secretory protein, the mRNA is translated in a heterologous protein-synthesizing system which lacks membranes. Such a system is one derived from wheat germ which is rich in ribosomes lacking their endogenous mRNA.

The mRNA in the cytoplasm associates with the membrane-free ribosomes to form polyribosomes. If the mRNA is for a secretory protein the first 20–25 amino acids from the N-terminus form a signal peptide which is rich in respect to hydrophobic amino acids and has a particular affinity for the membrane of the endoplasmic reticulum. The polyribosomes then become membrane-bound. If the mRNA is for a protein which is retained by the cell, a *housekeeping protein*, there is no signal peptide and the polyribosome does not attach to membrane.

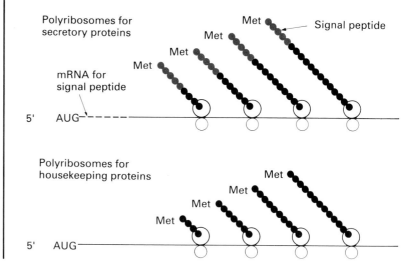

A

The signal hypothesis

In the original scheme, it was proposed that the signal peptide interacted directly with the membrane. It is now clear that the interaction is with the *signal recognition particle*. As the signal peptide emerges into the cisternae it is removed. Hence the mature protein or pro-protein is found in the membrane. The primary translation product is known as a preprotein, hence preproinsulin, preproparathyroid hormone, etc.

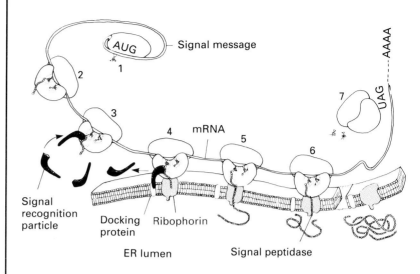

The role of various proteins has been elucidated. Thus the *signal recognition particle*, which consists of several different proteins and 7S RNA, binds to the signal peptide and arrests further translation until it locates the ribosome on the endoplasmic reticulum (ER) by the *docking protein*. Thus, secretory proteins are only made when the polysomes are bound to membrane. Various proteins contribute to the formation of the membrane pore, but their precise role is not yet clear. The signal recognition particle is released to the cytosol and translation by the membrane-bound ribosome is resumed.

B

Collagen synthesis

In the course of the biosynthesis of the insoluble protein collagen a precursor protein, procollagen, is synthesized which has extra peptides at each end.

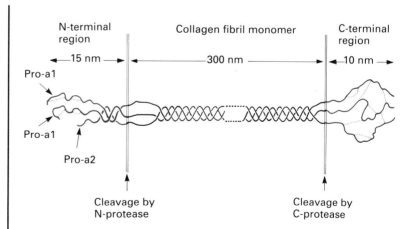

The extra peptides contain S–S bonds (indicated by ⋯⋯⋯) and serve to assist in the process of helical formation of the three chains of collagen, in this case shown as two identical chains (α1) and one different (α2). The peptides are lost either as the *procollagen* is secreted or just after secretion to give the insoluble protein *collagen*.

A

The formation of lysosomal enzymes

The *mucolipidoses.*

See also page 251.

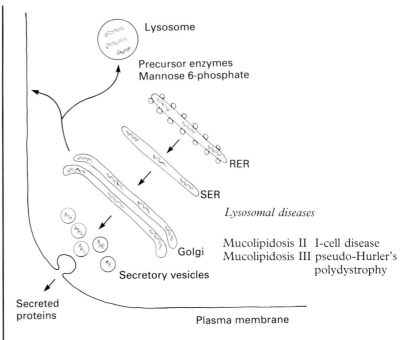

Lysosomal diseases

Mucolipidosis II I-cell disease
Mucolipidosis III pseudo-Hurler's polydystrophy

The lysosomal enzymes pass through the membranes as large precursor proteins and are glycosylated. Mannose residues are phosphorylated by a specific enzyme; an essential step in directing the enzyme to the lysosomes. It is this enzyme that is missing in the mucolipidoses in which the fibroblasts are characterized by their lack of lysosomal enzymes. Instead the enzymes are secreted extracellularly.

B

Synthesis of housekeeping proteins

The role of membrane-free polyribosomes in protein synthesis.

The synthesis of cytosolic, mitochondrial, *peroxisome,* and *chloroplast* proteins.

Nuclear proteins are synthesized in the cytoplasm and pass into the nucleus through the nuclear pore complex (page 2C) while RNAs are exported.

Apart from membrane-bound polyribosomes there exist in the cell membrane-free polyribosomes. These are the sites of translation of proteins for the cytosol, peroxisomes, mitochondria and, in plants, chloroplast proteins. The proteins may be made as larger precursors which are trimmed to mature proteins as they pass through the membranes. The process is *not* one of co-translational insertion.

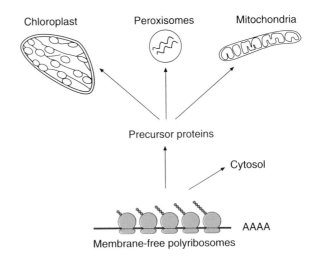

The origin of mitochondrial proteins

(In general terms the same applies to chloroplasts.)

Mitochondria in many respects resemble the prokaryotic bacteria and this has given rise to the suggestion that they arose in evolution by the incorporation of bacteria into eukaryotic cells.

In this regard they have a symbiotic existence, providing energy (ATP) for the cell but depending on the cytoplasm for newly synthesized proteins.

Mitochondrial DNA is derived solely from the mother since spermatozoa are devoid of mitochondria.

Mitochondria contain a small amount of circular DNA, tRNA, mRNA and ribosomes. As indicated previously (see p. 101 A), the ribosomes on which the mitochondrial proteins are synthesized resemble those of prokaryotes rather than eukaryotes (the same applies to chloroplasts.) The genetic code for mitochondria has some unique features (see p. 101 C) but in general resembles that of bacteria, e.g. initiation of translation by formyl-methionine (see p. 105 B). Most of the mitochondrial proteins are synthesized in the cytoplasm using the information in genomic DNA in the nucleus. Such proteins are transferred to the mitochondria. As shown in the figure there are four final destinations where the proteins are anchored. In most cases this is achieved by the synthesis of a protein which has additional peptides either at the N- or C-termini. These extra peptides serve as zip codes for the correct location of the proteins. There is some evidence that some proteins destined for location 3 or 4 are routed through the matrix. The chaperones (see p. 19) play a role in the transfer of the newly synthesized proteins through the outer mitochondrial membrane and in their anchorage in the matrix.

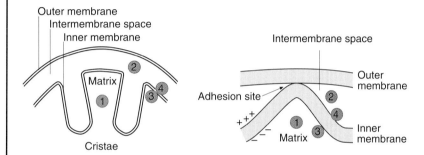

The locations of proteins synthesized in the cytoplasm within the mitochondria are shown above. The four sites are: 1, matrix; 2, intermembrane space; 3, attached to the inner membrane on the matrix side; 4, attached to the inner membrane on the outer side. The proteins pass to the inner membrane at the adhesion site.

A

Molecular variants of proteins

A summary showing their origin.

Molecular variants may arise normally by gene duplication from a single ancestral gene, e.g. the milk protein α-lactalbumin and enzyme lysozyme (see p. 254), or by alternative processing, e.g. proopiomelanocortin (see p. 40), or differential splicing of the primary product of the gene, e.g. antibody and in particular IgM$_{sec}$ and IgM$_{mem}$ (see pp. 46A, 122B).

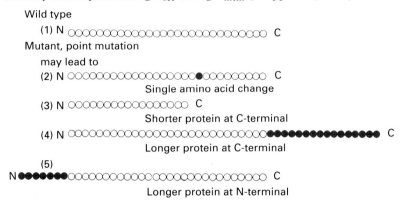

The *primary structure of molecular variants of proteins* is shown in the figure: 2, examples are the numerous haemoglobinopathies (see p. 27); 3, a mutation of a *sense* codon to a nonsense codon; 4, a mutation of a nonsense codon to a *sense* codon; 5, a mutation leading to replacement of a basic amino acid (Lys or Arg) so that cleavage of proprotein does not take place, e.g. proalbumin Christchurch, and insulinopathies.

B

Multiple steps in gene expression

Hn RNA = heterogeneous nuclear RNA see p. 121. mRNP = messenger ribonucleoprotein

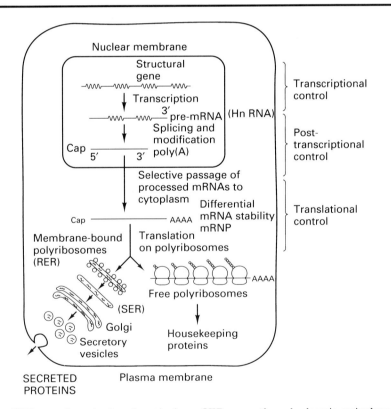

RER, rough endoplasmic reticulum; SER, smooth endoplasmic reticulum.

A

Protein degradation

If the cell is to contain no useless molecules then there must be processes whereby proteins and other macromolecules are eliminated when they are no longer required. In the case of the proteins most are subjected to 'turnover' (see p. 80 A). There are two routes for protein degradation: (1) the lysosomal system; (2) the ubiquitin system.

The lysosome system.

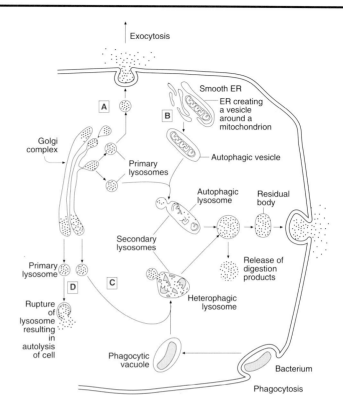

The formation of primary and secondary lysosomes and their role in cellular digestive processes is shown above. The primary lysosomes which bud from the Golgi body can take several pathways: A, exocytosis—transports enzymes to outside the cell; B and C, phagocytosis—formation of phagic lysosomes for digesting organelles or ingested matter; D, autolysis—destruction of the cell itself.

B

The ubiquitin system.

Ubiquitin is a 76 amino acid polypeptide that is present in virtually all types of cell, hence its name. Ubiquitin is activated by the interaction of ATP and two enzymes. It is this complex that then becomes attached to the protein targeted for degradation. Just how proteins are selected for destruction is not fully understood but there is a strong correlation between the nature of the N-terminal amino acid and those immediately following and the length of the survival time of a protein in a cell. Ubiquitin in yeast is synthesized as a pentamer which is then degraded to monomers.

The programmed destruction of cytosolic proteins by the ubiquitin-marking system is illustrated. E_2 is an enzyme involved in the ubiquitin transfer to lysine residues on target proteins.

6. RECOMBINANT DNA (GENETIC ENGINEERING)

A

Introduction

Plasmids are replicated together with chromosomal DNA.

An early demonstration of the transfer of DNA from one cell type to another concerned drug resistance by bacteria. The antibiotic resistance of certain strains of *Escherichia coli* resides in the extrachromosomal DNA of the *plasmid*. It was found that a plasmid could be transferred to a pathogenic bacterium, e.g. *Salmonella*, which then became resistant to the antibiotic. The plasmid contains the gene for the synthesis of an enzyme that is secreted by the bacteria and which destroys the antibiotic.

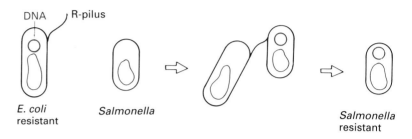

B

A first step in genetic manipulation is to insert a piece of foreign DNA into a vector (plasmid or virus) which then replicates in a suitable host. The inserted DNA can then be recovered by cleaving the vector. The recombinant DNA gene within the bacterium can be expressed in the form of protein. The host cells (e.g. bacteria) that contain the recombinant DNA are known as clones, so the process is referred to as 'cloning'.

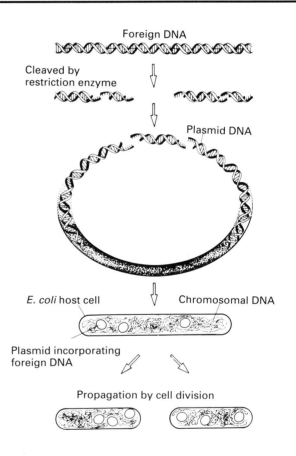

The methods used to insert the foreign DNA.

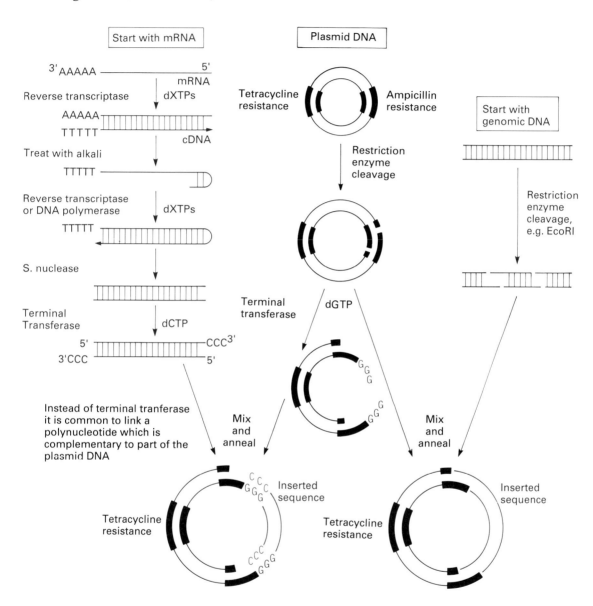

Two general methods for the construction of recombinant DNA molecules are illustrated. On the left, one starts with a partially purified mRNA and double-stranded complementary DNA (cDNA), with oligo(dC) tails, is formed. On the right, genomic DNA is degraded by a restriction enzyme into roughly 'gene-sized' pieces. The plasmid or vector has two antibiotic-resistant sites, one of which is cleaved by the same restriction enzyme as is used in the degradation of the DNA. The 'gene-sized' piece can be annealed into the cleaved plasmid DNA. In the other approach the cleaved DNA is tailed with oligo(dG) by means of terminal transferase and annealed to the cDNA. The host is transformed and cells selected for the appropriate insert. Antibiotic resistance is useful here, since, if the bacterial colonies are grown on a medium containing tetracycline, only those colonies (clones) that contain the plasmid will grow.

A

Restriction and other enzymes

The foreign DNA to be inserted into the *plasmid* is specially prepared by recombinant techniques.

The role of the three different DNA polymerases is described on page 95B.

The role of DNA ligase is described on pages 93 C and 95 B.

Various enzymes are used in the construction of recombinant DNA molecules. Enzymes 1, 2 and 3 are involved in synthesis and enzymes 4 and 5 in degradation of DNA. By means of enzyme 2 a new copy strand cDNA of the mRNA is formed. Enzyme 5 is particularly important. The *restriction enzymes* break DNA at very specific sites. Some 200 such enzymes have been identified.

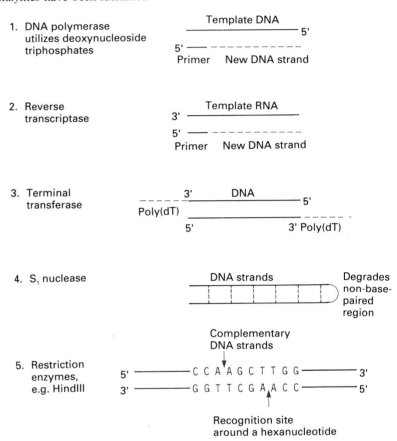

1. DNA polymerase utilizes deoxynucleoside triphosphates

2. Reverse transcriptase

3. Terminal transferase

4. S₁ nuclease

5. Restriction enzymes, e.g. HindIII

B

Nomenclature of restriction enzymes.

Restriction endonucleases are the means whereby a host bacterium protects itself from foreign DNA. An infecting phage is said to be 'restricted' by the host.

The restriction enzymes that recognize a particular target sequence in a duplex DNA molecule are known as type II enzymes. Several hundred such enzymes have now been at least partially characterized, having been isolated from a wide variety of bacteria. The nomenclature for these enzymes is as follows. The species name of the host organism is identified by the first letter of the genus name and the first two letters of the specific epithet to a three-letter abbreviation. Thus, *Escherichia coli* = *Eco* and *Haemophilus influenzae* = *Hin*. Strain or type identification is given by a further letter, hence *Hind*. Because of the symmetry of the recognition sequence the restriction enzyme may generate fragments with mutually cohesive termini as shown above (5). Other enzymes may generate fragments with flush ends.

A

Escherichia coli and alternative vectors

Although until recently *E. coli* and its plasmids have been the preferred vectors there is increasing interest in alternative systems. Thus, λ phage may be used as a vector in *E. coli*. This allows the use of larger inserts. *Bacillus subtilis* may be used as an alternative bacterium and yeast protoplasts with their own plasmids also show promise. Foreign DNA may be inserted into animal cells in culture either directly as a Ca^{2+} precipitate or incorporated into a virus. Such systems hold great promise where there are problems with secretory proteins.

B

Split genes, introns and exons

The transcription and translation of the ovalbumin gene.

The primary transcription product of the DNA is known as *heterogeneous nuclear RNA* (HnRNA) or pre-mRNA.

The molecular mechanisms for the removal of introns from pre-mRNA are shown on page 123.

An important application of recombinant DNA has been to examine the organization of the genomic DNA since the cDNA will hybridize with it. As indicated below, the bases which code for the egg-white protein ovalbumin are not continuous in the genome. This so-called split gene phenomenon is now commonly found, e.g. for haemoglobin α- and β-chains and other proteins. The intervening sequences in the genome are known as *introns* while the sequences which code for protein are known as *exons*.

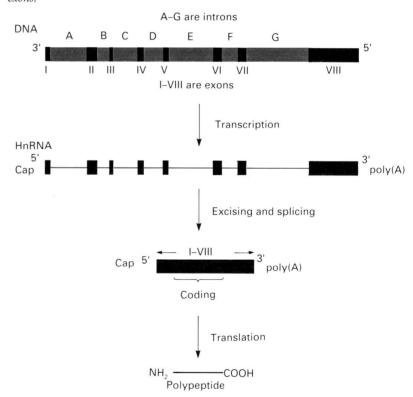

Most eukaryotic mRNA has a poly(A) tail at the 3' end added to the primary transcription product. The poly(A) tail is not translated.

A

Electron microscopy reveals the intervening sequences.

The mRNA for the chick protein *ovomucoid* was hybridized to the genome and the DNA/RNA hybrid visualized in the electron micrograph. The DNA is shown in black and the mRNA in red. The loops are lettered and show the lengths of DNA which do not hybridize to the mRNA and hence represent the intervening sequences.

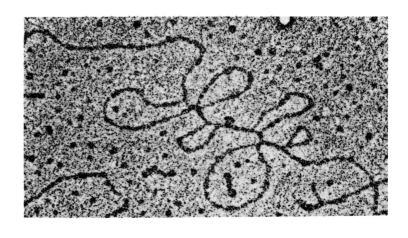

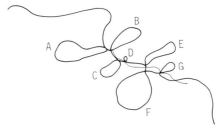

B

Alternative splicing

Splicing may be used for the generation of molecular variants of proteins. In the formation of IgM for secretion and membrane insertion (see also p. 45), different exons are selected for splicing to form a mature mRNA.

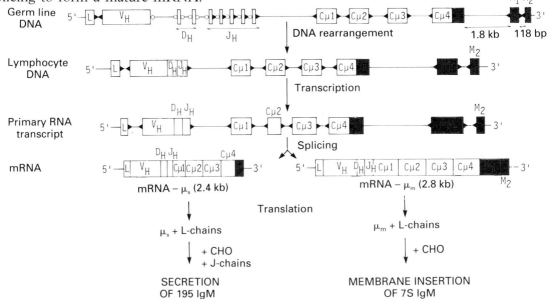

V_H, $C\mu1$, etc, represent the different domains of IgM shown on p. 43A. The red blocks indicate the exons that are distributed to the two mature mRNAs.

A

Mechanism of removal of introns

The structure of introns.

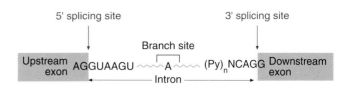

Splicing signals. Consensus sequences for the 5' splice site and the 3' splice site are shown.

It is important that the introns should be excised from the pre-mRNA very accurately and that the exons are correctly spliced together. The essential basis for this process is the common structural features of the introns. Thus, as shown above, at the 5' end of the introns of vertebrates there is *AGGUAAGU*, while at the 3' end there is a stretch of ten pyrimidines (U or C) followed by any base (N) and by C and ending with AG. There is also a site between 20 and 50 nucleotides upstream of the 3' splice site called the *branch site*.

B

Excising of introns and *splicing* of exons.

The figure below shows the events in the process and the utilization of the branch site for the formation of the 'lariat'. A group of *small nuclear ribonucleoproteins,* known as 'snurps', are involved in the process. The complex is known as a 'spliceosome'. This is a very large structure with an S-value of 40–60S. The role of the snurps was revealed by the use of antibodies obtained from patients with systemic lupus erythrematosus.

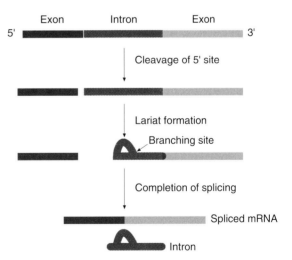

In the protozoan *Tetrahymena* the removal of the intron does not require proteins; indeed, an RNA is formed from the excised intron which has both nuclease and polymerase activities. This RNA is known as a *ribozyme.*

A

The use of radioactive probes to detect gene sequences

The *Southern blot.*

A similar method for RNA is known as the *Northern blot.*

A similar method for proteins using antibodies for detection is known as the *Western blot.*

The Southern blot technique

In order to detect fragments of DNA in an agarose gel that are complementary to a given RNA or DNA sequence, Southern devised a method for transferring denatured DNA to cellulose nitrate (nitrocellulose). The DNA fragments can be permanently fixed to the cellulose nitrate by heating at 80°C. Subsequently, a radioactive probe can be hybridized to the cDNA on the cellulose nitrate.

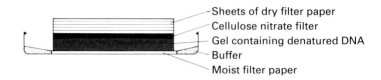

The dry filter paper draws the buffer solution up from the gel, carrying the DNA with it into the cellulose nitrate.

B

Mapping restriction sites around a hypothetical gene sequence in total genomic DNA.

Radioactive probes can also be used on tissue sections to identify the site of a nucleic acid. This is called 'in-vitro hybridization'.

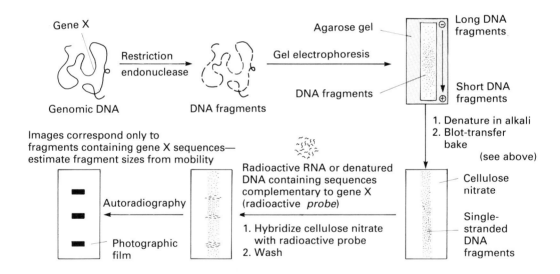

For the preparation of the radioactive *probe* it is necessary to know the base sequence of at least a part of gene X.

Techniques for the determination of the base sequence of DNA

Maxam and Gilbert sequencing procedure.

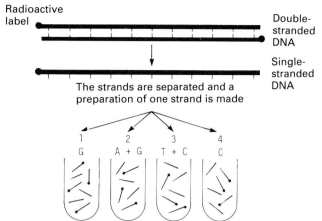

The strands are separated and a preparation of one strand is made

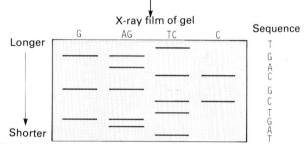

A chemical agent destroys one or two of the four bases and so cleaves the strands at those sites. The reaction is controlled so that only some strands are cleaved at each site, generating a set of fragments of different sizes

The fragments are separated according to size by gel electrophoresis and the radioactive fragments produce images on an X-ray film. The images on the X-ray film determine which base was destroyed to produce each radioactive fragment

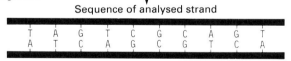

Sequence of analysed strand

Sequence of complementary strand

The sequence of the destroyed bases yields the base sequence of the analysed strand

A segment of DNA is labelled at one end with [32]P. The labelled DNA divided into four samples is treated with a chemical that specifically destroys one or two of the four bases. Only a few sites are nicked in any one DNA molecule. When the nicked molecules are treated with piperidine the fragments break up. A series of labelled fragments is generated, the length of which depend on the distance of the destroyed base from the labelled end of the molecule. The sets of labelled fragments are run side by side on an acrylamide gel that separates DNA fragments according to size and the gel is autoradiographed. The pattern of bands on the X-ray film is read to determine the base sequence of DNA.

Sanger sequencing
procedure.

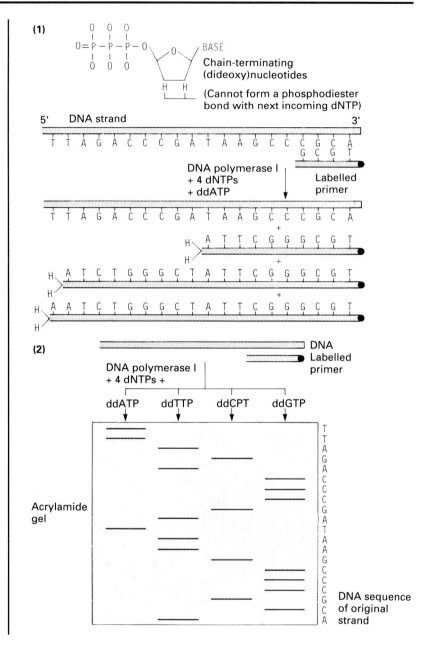

1. In the Sanger procedure, 2',3'-dideoxynucleotides of each of the four bases are prepared. These molecules are incorporated into DNA by *Escherichia coli* DNA polymerase but, once incorporated, the dideoxynucleotide (ddNTP) cannot form a phosphodiester bond with the next incoming dNTP and so the growth of the chain stops. A sequencing reaction consists of a DNA strand to be sequenced, a short labelled piece of DNA (the primer) that is complementary to the end of that strand, a carefully controlled ratio of one particular ddNTP with its normal dNTP and the three other dNTPs. If the correct ratio of ddNTP to dNTP is chosen, a series of labelled strands will result, the lengths of which depend on the location of a particular base relative to the end of the DNA.

2. A DNA strand to be sequenced, along with labelled primer, is split into four DNA polymerase reactions, each containing one of the four ddNTPs. The resultant labelled fragments are separated by size as for the Maxam and Gilbert procedure.

A

Antenatal diagnosis

An example of the role of recombinant DNA technology in antenatal diagnosis.

A particular application has been to devise means whereby it is possible to determine the genetic make-up of the fetus being carried by the mother. With certain knowledge of the status of the fetus the mother may in some societies be able to choose to have an abortion. An example is the problem of sickle-cell anaemia and methods to determine whether the fetus is homozygous for the sickle-cell trait. The genetic change is located in the β-chain of HbA. As shown below, the fetus typically contains HbF, but after a few weeks there is some HbA, which may be analysed. Blood is withdrawn by a process known as fetoscopy, which carries some dangers.

B

Haemoglobin polymorphism in the human.

The diagram shows how the different types of haemoglobin present in the normal human arise from the gene loci.

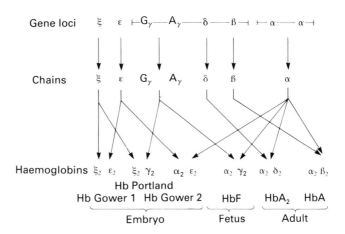

C

The analysis of the genome.

The use of a *gene probe* for the detection of sickle-cell disease.

Scattered throughout the human genome there are harmless base changes which either produce new restriction endonuclease sites or remove pre-existing ones. The changes which are inherited are called *restriction fragment length polymorphisms* (RFLPs). These are often linked to genes related to disease as in the case illustrated.

The association of an RFLP with the HbS mutation

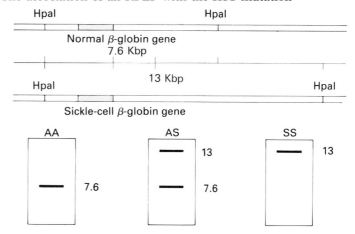

Illustrated is an indirect method which depends on a mutation in the non-coding DNA which is often linked to the point mutation leading to HbS.

In this technique, as an alternative to examination of fetal haemoglobin, some amniotic fluid is withdrawn (amniocentesis) which contains some of the fibroblast cells from the embryo. The DNA of the fibroblast is treated with a restriction enzyme (HpaI). The fragments are separated by the Southern blot technique and probed with a [^{32}P]cDNA for β-globin. The size of the fragments (shown in kilobases) differs in HbA and HbS. The change in the restriction site is frequently associated with the sickle-cell mutation in certain populations.

A

The diagnosis of sickle-cell anaemia by restriction enzyme analysis of fetal DNA.

Instead of the use of RFLPs, the presence of the HbS gene may be detected directly.

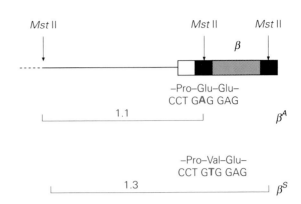

The base substitution which causes HbS removes an *Mst*II site. In normal DNA, an *Mst*II fragment of 1.1 kb is generated. The absence of the *Mst*II site in those with HbS generates a longer *Mst*II fragment of 1.3 kb. Heterozygotes show both fragments, homozygotes only one.

B

Reverse genetics

The strategy is to correlate the phenotype for a disease with a chromosome deletion. The gene is then sequenced and the amino acid sequence of the protein deduced. The function of the protein is then predicted.

This strategy is also termed 'positional cloning' to differentiate it from 'functional cloning' where the role of the aberrant gene is known.

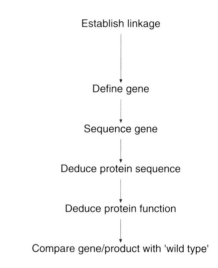

A success of the above strategy is the elucidation of the genetic basis of cystic fibrosis. By the use of standard linkage markers the gene was located on chromosome 7. A candidate gene was located and sequenced. The gene codes for the cystic fibrosis transmembrane conductance regulator (CFTR), a chloride channel requiring both cAMP-dependent phosphorylation and ATP hydrolysis to open.

The cause of Duchenne muscular dystrophy (DMD) has also been studied. The gene involved was located on the X-chromosome and codes for a protein of M_r 400 000 named 'dystrophin'. The precise function of this protein is not at present clear.

The polymerase chain reaction (PCR)

The PCR has very wide applications, particularly in medical diagnosis where only small amounts of starting DNA are available. By means of PCR, 10- to 100 000-fold amplification of the DNA stretch required may be obtained.

With a knowledge of the sequence of bases that flank the region of the DNA to be amplified, oligonucleotides complementary to these sequences are prepared. The DNA containing the sequences to be amplified is heat denatured (step 1) and annealed to the primers (step 2). Polymerase chain extension is carried out from the primer termini (step 3). Subsequent cycles of denaturation, annealing and primer extension are carried out. By using a thermophilic DNA polymerase (*Taq*) which is not destroyed at the temperature of the heat denaturation the addition of enzyme at each cycle is avoided. Apparatus is available to automate the recycling process.

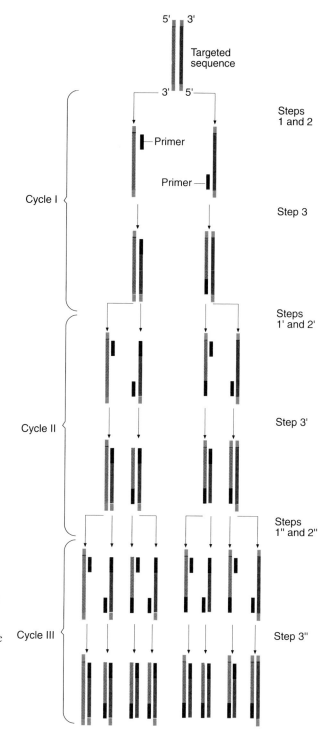

A brief glossary of terms not previously explained

'Alu' family DNA
The most abundant family of repetitive sequences in mammals, present throughout the genome. The human Alu sequence consists of 300 bp and appears about 300 000 times. The Alu sequence is released by the action of the Alu restriction enzyme, hence its name.

Amplification
1. Treatment designed to increase the proportion of plasmid DNA relative to that of bacterial DNA. 2. Replication of a gene library in bulk.

Cap
The structure found at the 5' end of many eukaryotic mRNAs; it consists of 7'-methyl-guanosine-pppX, where X is the first nucleotide encoded in the DNA; it is not present in prokaryotic mRNAs.

Cistron
A DNA fragment or portion that specifies or codes for a particular polypeptide.

Cosmid cloning
A technique for cloning large eukaryotic fragments in *Escherichia coli*. It employs the single-stranded 'cos' sites at each end of λ phage which are required for packaging in phage heads.

Episome
A circular gene fragment.

Frame shift
A mutation that is caused by insertion or deletion of one or more paired nucleotides, and whose effect is to change the reading frame of codons during protein synthesis, thus yielding a different amino acid sequence beginning at the mutated codon.

Fusion proteins
Hybrid proteins containing both bacterial and eukaryotic amino acid sequences.

Genome
All the genes of an organism or individual.

Grünstein–Hogness assay colony
Hybridization procedure for identification of plasmid clones (colonies are transferred to a filter and hybridized with a probe).

Heat shock genes
High temperatures and other stress-inducing treatments evoke the expression of heat shock genes in yeasts and other eukaryotes giving rise to heat shock proteins.

Heteroduplex
A DNA molecule, the two strands of which come from different individuals so that there may be some base pairs or blocks of base pairs that do not match.

Inversion
The alteration of a DNA molecule made by removing a fragment, reversing its orientation, and putting it back into place.

Klenow polymerase
DNA polymerase I possesses exonuclease activity in the absence of dNTPs and will degrade single strands of DNA duplex in both directions. Removal of part of the polypeptide chain to produce the Klenow fragment leaves only 3'→5' exonuclease activity.

Linker
A small fragment of synthetic DNA that has a restriction site useful for gene splicing.

M13 phage
A single-stranded DNA phage. The double-stranded replicative form can be used as a cloning vector.

Metallothionine (MMT)
A metal-binding protein that is induced in animal cells by a variety of metals, e.g. zinc. Using the MMT gene promoter fused to the coding region of the protein required, the expression of a fusion protein can be regulated.

Nick translation procedure
Procedure for labelling DNA in vitro using DNA polymerase I.

Open-reading frames
Long stretches of triplet codons in DNA that are not interrupted by a translational stop codon.

Palindrome
A self-complementary nucleic acid sequence, i.e. a sequence identical to its complementary strand (both read in the same 5' to 3' direction).

Restriction maps
The location of the multiple sites within a DNA which are susceptible to cleavage by a variety of restriction enzymes. Such a map will indicate the degree of homology of different DNA molecules.

Shine–Dalgarno sequences
A ribosome-binding site in mRNA about 8 nucleotides upstream from the initiation codon which is closely complementary to the 3' end of the smaller (16S) of the two rRNA molecules in bacteria.

Transduction
The transfer of genetic material from one cell to another by means of a viral vector (for bacteria, the vector is bacteriophage).

Transfection
Infection of a cell with isolated DNA or RNA from a virus or viral vector.

Transformation
The introduction of an exogenous DNA preparation (transforming agent) into a cell.

Transgenic
Animals that have integrated foreign DNA into their germ line as a consequence of experimental introduction of DNA, commonly by microinjection or retroviral infection.

Yeast artificial chromosomes (YACS)
A cloning vector formed from bacterial plasmids, two yeast telomeres (chromosome tips), a yeast centromere and other elements which allow for the accommodation of human DNA fragments. The vector can be introduced into yeast, where it functions as an artificial chromosome.

COENZYMES AND WATER-SOLUBLE VITAMINS

Early attempts to purify enzymes led to the discovery of compounds that were lost on dialysis, which were essential for enzyme activity. These compounds became known as *coenzymes*. The term *cofactor* includes such compounds, but also includes other molecules such as metal ions that may be necessary for enzyme activity, whereas the term 'coenzyme' is reserved generally for organic molecules participating in the reaction. Indeed, a number of coenzymes, such as the nicotinamide nucleotides, are included in the written reaction, appearing in their transformed form as products. Kinetically, they are treated as a substrate in their own right. Other coenzymes, such as pyridoxal phosphate, are not normally shown as substrates, even though they may undergo transformation in the course of the reaction. In some cases, the coenzyme may be tightly bound to the enzyme. This is true, for example, in the case of flavin nucleotides such as FAD, which is reduced and reoxidized while bound to the enzyme protein.

It has been found that water-soluble vitamins are involved in providing components of these major coenzymes.

A

Introduction

As enzymes came to be obtained in a highly purified form, it was found that their activity often depended on the presence of certain small molecules that came to be known as coenzymes. Coenzymes are molecules that cooperate in the catalytic action of an enzyme. A coenzyme may be tightly bound to the protein, or it may in other cases be free to diffuse away from the enzyme, acting essentially as an additional substrate (e.g. NAD^+ in many dehydrogenase reactions). If tightly bound to the enzyme, it may sometimes be referred to as a *prosthetic group*. The term *cofactor* refers rather non-specifically to the various species required for full enzyme activity, including cations (e.g. Zn^{2+} or Mg^{2+}). An enzyme protein from which the coenzyme has been removed is referred to as an apo-enzyme. Many water-soluble vitamins were found to be coenzymes, or components of coenzyme molecules. Compounds of differing structures that have the same vitamin action are known as *vitamers*.

B

Nicotinamide nucleotides

Nicotinamide adenine dinucleotide (oxidized form) (NAD^+) and nicotinamide adenine dinucleotide phosphate (oxidized form) ($NADP^+$), and their reduced forms (NADH and NADPH), are involved in a great many dehydrogenase reactions in the mitochondrion, cytosol and endoplasmic reticulum of the cell. They are water-soluble, and are usually free to diffuse away from the enzyme, after conversion to the oxidized or reduced form, to take part in another dehydrogenase reaction catalysed by another enzyme.

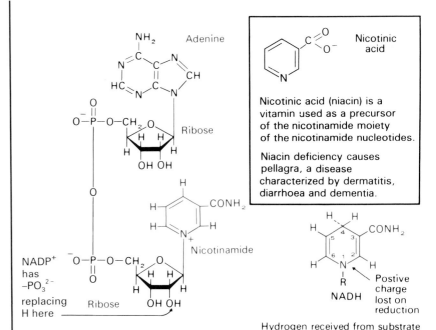

Nicotinic acid (niacin) is a vitamin used as a precursor of the nicotinamide moiety of the nicotinamide nucleotides.

Niacin deficiency causes pellagra, a disease characterized by dermatitis, diarrhoea and dementia.

Hydrogen received from substrate is bound to the nicotinamide moiety.

C

The spectrum of NADH differs from that of NAD^+ in that there is a peak of absorption around 340 nm.

Many dehydrogenase reactions can be followed by measuring the change in absorbance at 340 nm.

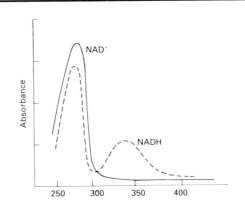

A

Adenine nucleotide coenzymes

The ultimate purpose of tissue respiration is the phosphorylation of ADP to produce ATP. The overall direction of cellular metabolism is regulated by this process.

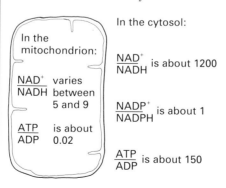

ATP

Adenosine 5'-phosphate = adenosine monophosphate (AMP)
Adenosine 5'-diphosphate = adenosine diphosphate (ADP)
Adenosine 5'-triphosphate = adenosine triphosphate (ATP)

ATP is a coenzyme of very great importance in cell metabolism. The large negative free energy change that is characteristic of reactions in which it participates is responsible for driving many reactions in the direction leading to ATP hydrolysis, or phosphoryl group transfer. In addition, the ATP/ADP/AMP system is involved in cellular regulation through allosteric mechanisms. Many examples of these actions will be found in subsequent chapters.

The prefixes 'di-' and 'tri-' are used for compounds in which the phosphate radicals are in anhydride linkage with each other. A compound which has more than one phosphate radical at different sites is described as a bis- or trisphosphate.

B

The relative proportions of the components of the nicotinamide and adenine nucleotide coenzyme systems differ between the cytosolic and mitochondrial compartments of the cell. This has a profound effect on the direction of metabolism in these two compartments.

Some enzymes have tightly bound NAD^+ that acts as a modulator of the protein conformation. These coenzymes are allosteric effectors of many enzymes. In addition, they influence enzyme activity by virtue of their relative concentrations by a mass action effect.

In the mitochondrion:

$\dfrac{NAD^+}{NADH}$ varies between 5 and 9

$\dfrac{ATP}{ADP}$ is about 0.02

In the cytosol:

$\dfrac{NAD^+}{NADH}$ is about 1200

$\dfrac{NADP^+}{NADPH}$ is about 1

$\dfrac{ATP}{ADP}$ is about 150

Thus in the cytosol the tendency is for reactions involving NAD^+ to proceed towards NADH (oxidation of the substrate). Because of this, NADPH is often the coenzyme used for reduction in the cytosol, which the more favourable ratio of reducing and oxidizing equivalents (redox ratio) facilitates.

The general form of reactions involving NAD^+ is

$$RH_2 + NAD^+ \rightleftharpoons R + NADH + H^+$$

For convenience, in writing metabolic pathways, the H^+ is omitted. However, it should be appreciated that it is always involved in the reaction.

A

Adenine nucleotide interrelationships.

The enzyme myokinase interconverts the adenine nucleotides.

$$ATP + AMP \underset{\text{Myokinase}}{\rightleftharpoons} 2ADP$$

The metabolic status of a cell is indicated by the relative concentrations of the adenine nucleotides. A useful index is the *energy charge*, defined as

$$\frac{[ATP] + \frac{1}{2}[ADP]}{[ATP] + [ADP] + [AMP]}$$

B

Thiamin pyrophosphate

Thiamin (vitamin B_1) is the precursor of thiamin pyrophosphate, the coenzyme for some important oxidative decarboxylation reactions (pyruvate dehydrogenase, oxoglutarate dehydrogenase). A deficiency of thiamin causes the disease beriberi, a disease associated with neuropathy and cardiopathy. Experimental deficiency causes neurological symptoms (pigeons fail to hold their head erect) that can readily be reversed by administering the vitamin. In oxidative decarboxylation, loss of CO_2 is accompanied by oxidation of an aldehyde to an acid.

Thiamin pyrophosphate is important in the pyruvate dehydrogenase and similar reactions.

$$Pyruvate + CoA \xrightarrow{\quad NAD^+ \quad NADH \quad} Acetyl\ CoA + CO_2$$

Pyruvate dehydrogenase reaction

The pyruvate dehydrogenase enzyme has a complex mechanism involving binding sites for pyruvate, NAD^+, thiamin pyrophosphate and lipoic acid.

A

Flavin nucleotide coenzymes

FAD is the coenzyme of a class of dehydrogenases known as flavoproteins. The flavin moiety of the molecule is derived from riboflavin (vitamin B_2).

Flavin mononucleotide (FMN) is an important coenzyme in some flavoproteins, including NADH coenzyme Q reductase (see p. 182).

FMN consists of riboflavin phosphate (i.e. FAD without the adenosine monophosphate moiety).

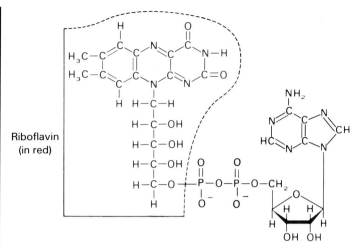

Flavin adenine dinucleotide (FAD)

Reduction of FAD or FMN involves the two unsubstituted N atoms of the isoalloxazine structure, as shown in red above.

B

Flavin coenzymes remain tightly bound to the enzyme protein throughout the reaction, in contrast to nicotinamide coenzymes that bind reversibly to the enzyme.

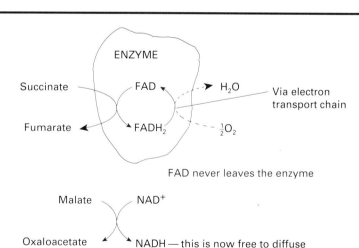

A number of mitochondrial dehydrogenases are flavoproteins; for example, NADH dehydrogenase, succinate dehydrogenase and fatty acyl CoA dehydrogenase.

A

Coenzyme A

Coenzyme A (CoA) is a complex molecule which contains a free sulphydryl (– SH) group. This group can react with a carboxyl group to form a thioester.

Pantothenate is an essential food factor which forms part of CoA. The free sulphydryl group can react with a carboxyl group to form a thiol ester = thioester. Such thiol esters are involved in many transfer reactions involving acyl groups, including acetyl and fatty acyl groups.

A thioester is in some ways analogous to the ester formed between a carboxylic acid and an alcohol. However, it is considerably more reactive. In acetyl CoA, the thioester linkage can activate the methyl carbon as well as the acyl carbon.

Coenzyme A and acetyl coenzyme A

B

Biotin

Biotin is an essential food factor and is a coenzyme for certain carboxylation reactions (e.g. pyruvate carboxylase). These reactions involve ATP, which is necessary in the first step of the reaction; this is the conversion of biotin to N-carboxybiotin by the addition of CO_2. Biotin is covalently bound to the enzyme by its carboxyl group.

A

Folate coenzymes

The transfer of a single methylene or methyl group often involves folic acid, a water-soluble vitamin, in one of its several substituted forms.

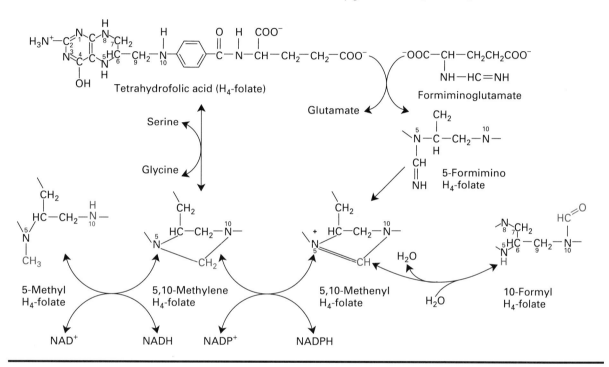

B

The different folate coenzymes are specific for particular reactions.

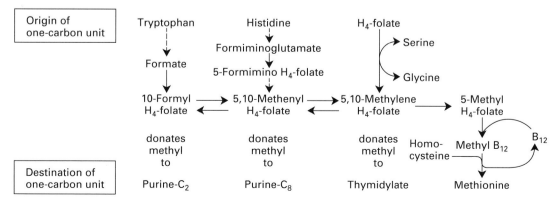

Pyridoxal phosphate

Pyridoxal phosphate (derivative of vitamin B_6) acts as coenzyme in transamination and decarboxylation reactions. In a transamination reaction the aldehyde group of pyridoxal phosphate first forms a Schiff base with the amino group of the amino acid, which is then converted to keto acid. Pyridoxal phosphate is thereby converted to pyridoxamine phosphate, which transfers the amino group to another keto acid to form an amino acid.

Pyridoxal phosphate

Pyridoxine (Vitamin B_6 derivative)

$$R-CH-COOH \atop +\ NH_2 \atop X-CH{=}O \quad \xrightarrow{-H_2O} \quad R-CH-COOH \atop N \atop X-C-H \quad \longrightarrow \quad R-C-COOH \atop N \atop X-CH_2 \quad \xrightarrow{+H_2O} \quad R-C-COOH \atop O \atop +\ NH_2 \atop X-CH_2$$

Amino acid A + Pyridoxal-P   ·   Schiff bases   ·   Keto-acid A + Pyridoxamine-P

$$R_1-C-COOH \atop O \atop NH_2 \atop X-CH_2 \quad \xrightarrow{-H_2O} \quad R_1-C-COOH \atop N \atop X-CH_2 \quad \longrightarrow \quad R_1-CH-COOH \atop N \atop X-CH \quad \xrightarrow{+H_2O} \quad R_1-CH-COOH \atop +\ NH_2 \atop X-CH{=}O$$

Keto-acid B + Pyridoxamine-P   ·   Schiff bases   ·   Amino acid B + Pyridoxal-P

$$X =$$ (pyridoxal-P ring structure: HO, CH_2OP, H_3C, N)

Pyridoxal phosphate also acts as a coenzyme in decarboxylation reactions of amino acids such as

$$\begin{array}{c} COO^- \\ | \\ CH_2 \\ | \\ CH_2 \\ + | \\ H_3NCHCOO^- \end{array} \quad \xrightarrow[CO_2]{H^+} \quad \begin{array}{c} COO^- \\ | \\ CH_2 \\ | \\ CH_2 \\ + | \\ H_3NCH_2 \end{array}$$

Glutamate

γ-Aminobutyrate (GABA, a neurotransmitter)

Vitamin B_6 deficiency is rare in humans, because the vitamin is widely distributed in common foodstuffs and, in addition, is synthesized in appreciable quantities by intestinal flora. The main abnormality seen in B_6 deficiency is a dermatitis, which is readily cured by administration of the vitamin. In addition, however, laboratory animals made B_6-deficient may be susceptible to spontaneous epileptiform convulsions, and are more susceptible to convulsive drugs, which act as GABA antagonists.

Vitamin B_6 is a general term for pyridoxine, pyridoxal and pyridoxamine.

A

Vitamin B$_{12}$

Vitamin B$_{12}$ was first isolated as the factor in raw liver that alleviates the symptoms of pernicious anaemia. Four pyrrole rings form a porphyrin-like structure around a cobalt atom. The vitamin acts as a coenzyme in the methylation of homocysteine, and in a reaction in which methylmalonyl CoA is converted to succinyl CoA. Both reactions involve a methyl group transfer.

Other forms exist. In cyanocobalamin (the form first isolated) a –CN group occupies the sixth coordination position in place of the 5′-deoxy adenosine moiety (shown in red); in methylcobalamin this position is occupied by a methyl group.

Structure of the 5,6-dimethylbenzimidazole cobamide coenzyme.

B

Ascorbic acid (vitamin C)

The action of ascorbic acid in preventing poor wound healing in scurvy may be explained by its involvement in the hydroxylation of proline.

Ascorbic acid is found in fresh vegetables and fruit, and a deficiency causes scurvy, a disease characterized by defective connective tissue, with bleeding gums and loss of teeth.

Proline is hydroxylated to hydroxyproline after being incorporated into collagen precursor peptides (see p. 38 and 113B):

—Pro—Gly— + α-Oxoglutarate + O$_2$ →

    —Hypro—Gly— + CO$_2$ + succinate + H$_2$O

Fe^{3+} and ascorbic acid appear to be required in this reaction. In the absence of ascorbic acid an abnormal collagen is formed.

Ascorbic acid undergoes ionization and can act as a reducing agent, being oxidized to dehydroascorbate.

Ascorbate   Ascorbic acid   Dehydroascorbic acid

6

CARBO-HYDRATE CHEMISTRY AND INTER-CONVERSIONS OF MONO-SACCHARIDES

Carbohydrates may be considered as having two distinct and vital functions, first as substrates, the oxidation of which provides energy for synthetic reactions and work, and second as structural elements.

In relation to the first function, glucose is of special importance. The energy-storage compounds starch, in plants, and glycogen, in animals, are polymers of glucose. It acts as the substrate for glycolysis, one of the main energy-producing oxidative pathways.

In the structural role, various sugars apart from glucose, including mannose, galactose and fucose and their aminated and sulphated derivatives, are present in complex polysaccharides, which play an important part in recognition functions and adhesion, as explained in later sections.

Sugars are classified as hexoses (if six-carbon compounds) or pentoses (five carbons). They are aldoses if the carbonyl group is on a terminal carbon, or ketoses if the carbonyl group is a ketone. They form groups of isomeric molecules, distinguished from each other by the configuration of their hydroxyl groups. Because each sugar has a number of free hydroxyl groups, any of which may be involved in bonding to other sugars, the number of possible polymers that may be derived from a small number of different monomers is very great.

A

Chirality of sugars

D-forms of the monosaccharides predominate in mammalian carbohydrates.

Simple sugars contain centres of asymmetry and are thus chiral molecules (see p. 10). Every sugar exists in a D- or L-form, each being a mirror image of the sugar.

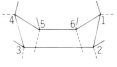

CHO	1	CHO
H—C—OH	2	HO—C—H
HO—C—H	3	H—C—OH
H—C—OH	4	HO—C—H
H—C—OH	5	HO—C—H
CH₂OH	6	CH₂OH

D-Glucose   L-Glucose

The configuration on carbon 5 determines whether the sugar is D or L (the relationship can be traced back to D- or L-glyceraldehyde). Since these two sugars are mirror images, the configuration of all the other hydroxyls is reversed also in the Fischer projection shown on the left.

Note: The configuration at each carbon can also be denoted *R* or *S* (see p. 10). This notation applies to individual carbons, not the whole molecule.

B

Monosaccharides normally adopt a ring conformation.

A six-membered ring structure may adopt either a chair or boat conformation, shown below for cyclohexane. Bonds perpendicular to the plane of the molecule are termed axial, whilst those in the plane of the molecule are equatorial.

Boat   Chair   Equatorial bonds   Axial bonds

When D-glucose adopts a six-membered ring structure a *hemi-acetal* bond is formed between the aldehyde carbonyl group of C-1 and the hydroxyl group on C-5.

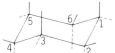   + HOR

Aldehyde   Alcohol   Hemi-acetal

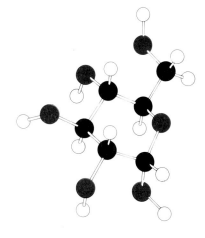

If a further alcohol group reacts, a full acetal  is formed (this type of structure is found in glycosides, see p. 143).

The six-membered ring of D-glucose adopts a chair configuration.

The hemi-acetal group (but not a full acetal) reacts readily with certain oxidizing agents such as Fehling's solution and this forms the basis of the well-known 'reducing test' for sugars with a hemi-acetal group, such as glucose or mannose. Some disaccharides (see p. 144) possess one hemi-acetal group, but sucrose has no hemi-acetal group, so does not react with Fehling's solution.

A

Monosaccharides

Carbohydrate structures.

Simple sugars are termed monosaccharides.
The α- and β-forms of glucose spontaneously undergo interconversion *(mutarotation)* in water.

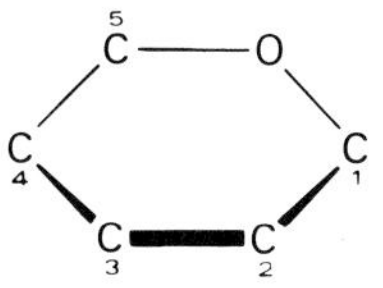

Numbering for a monosaccharide in the pyranose form

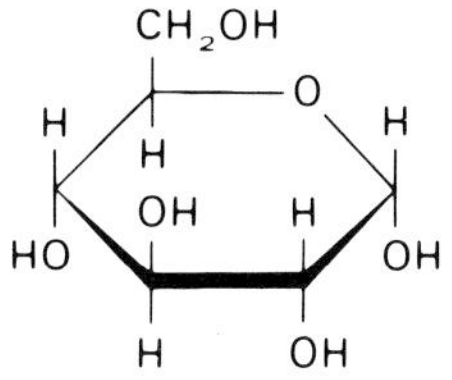

α-D-Glucose

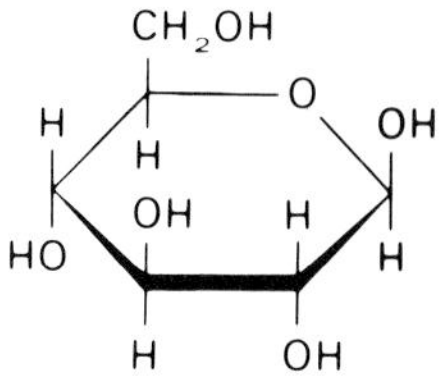

β-D-Glucose

Two conformations of D-glucose

The *pyranose* form of a sugar has a six-membered ring, while the *furanose* form has a five-membered ring.

α-D-Glucose and β-D-glucose differ only in the conformation of the hydroxyl group on carbon 1. These and other pairs of sugar molecules that differ only in this respect are termed *anomers*.

B

The ring form of the carbohydrate is in equilibrium with an open-chain form.

The open-chain form aids understanding of mutarotation, and of the difference between glucose, which has an aldehyde group on C-1, and is thus called an aldose, and fructose, which has a ketone group on C-2, and is called a ketose sugar. A sugar with six carbons is a hexose, one with five carbons is a pentose.

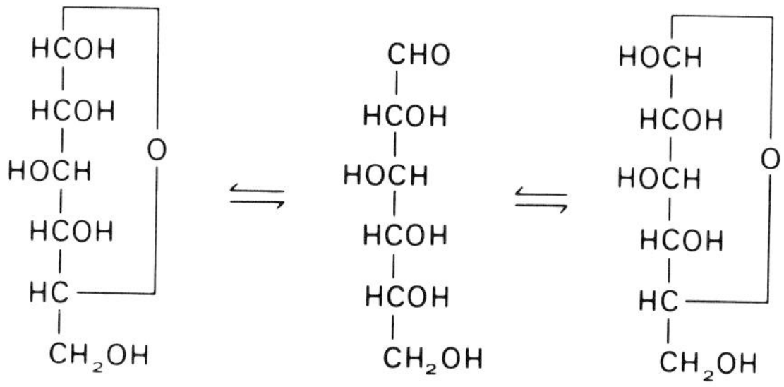

α-D-Glucopyranose

β-D-Glucopyranose

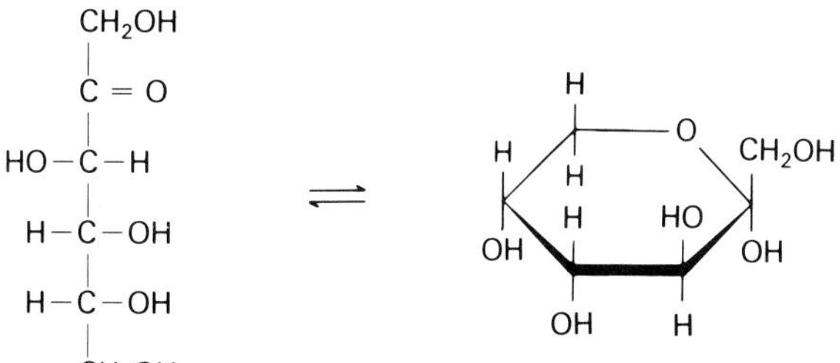

α-D-Fructopyranose

C

Enzymes that bring about reactions at C-1 between the sugar and an alcohol yield a glycoside that has either the α- or the β-configuration.

Note: Methyl glucoside is a glycoside of glucose.

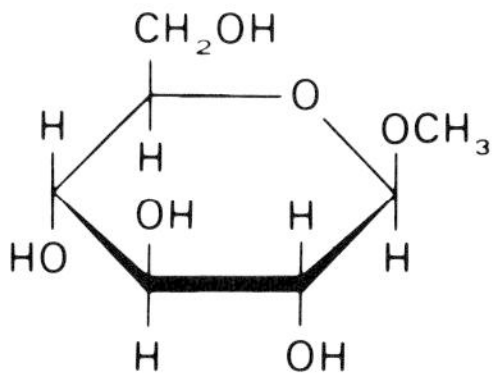

Methyl β-D-glucoside

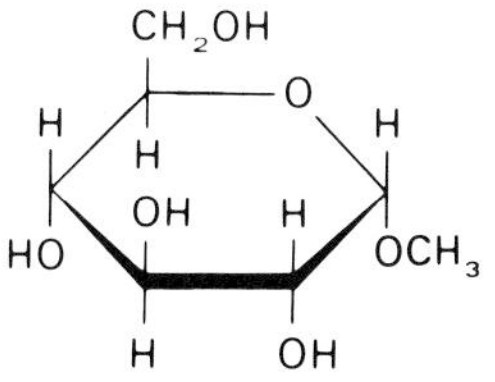

Methyl α-D-glucoside

There is no mutarotation of these compounds. Enzymes that act on glycosides are specific for the α- or β-form, and are termed *glycosidases*.

A

Other monosaccharides (simple sugars) differ from one another by virtue of the configuration of one of the optically active hydroxyl groups.

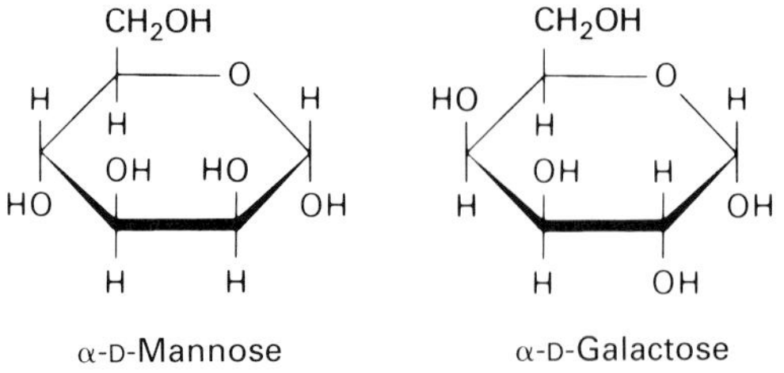

α-D-Mannose

α-D-Galactose

Isomers of this type are termed *epimers*, and enzymes that change the configuration on one hydroxyl group to convert one sugar into another are termed *epimerases*.

B

Important pentoses are ribose and xylulose.

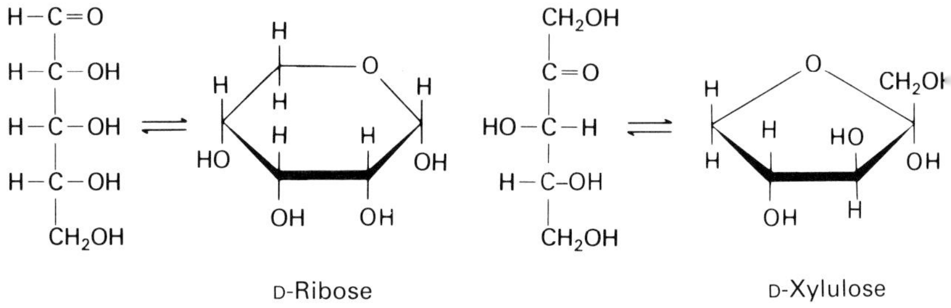

D-Ribose

D-Xylulose

C

Amino sugars have a nitrogen on C-2 that is often acetylated.

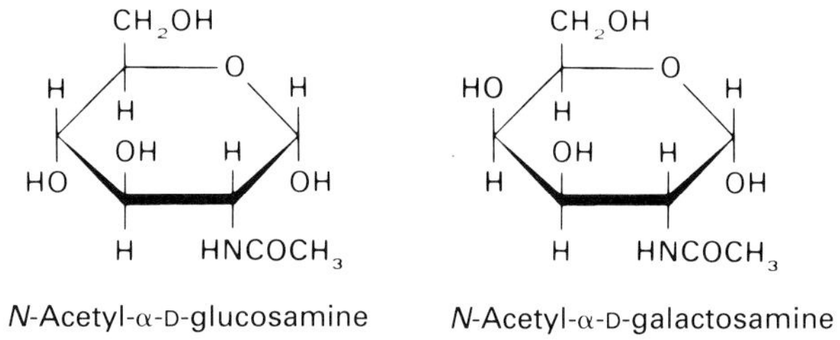

N-Acetyl-α-D-glucosamine

N-Acetyl-α-D-galactosamine

D

Disaccharides

Maltose is a *disaccharide* formed from two molecules of glucose, linked between C-1 of one glucose and C-4 of the other. The link at C-1 is in the α- configuration, so that an α-(1→4) link results.

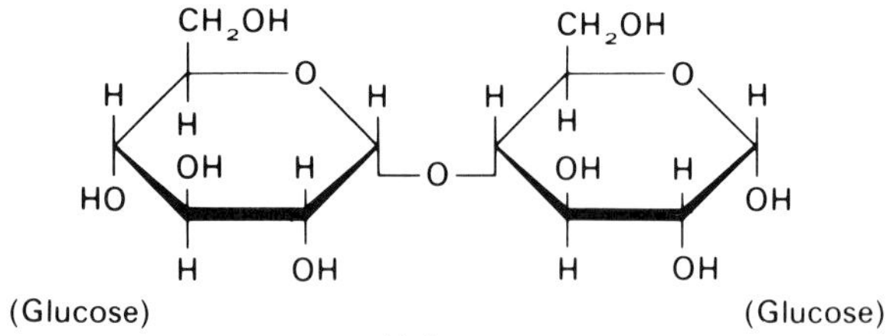

(Glucose)

(Glucose)

Maltose

(α-D-Glucosyl-(1 → 4)-α-D-glucose)

A

Sucrose and lactose have β-(2→1) and β-(1→4) links, respectively.

The carbonyl group on C-1 of monosaccharides is termed a reducing group as it can be oxidized readily by weak oxidizing agents such as alkaline copper sulphate. This does not occur if C-1 is engaged in bonding to another compound. Thus, sucrose has no free reducing group.

Lactose is the sugar found in milk.

(Glucose)   (Fructose)

Sucrose

(β-D-fructosyl-α-D-glucose)

(Galactose)   (Glucose)

Lactose

(β-D-galactosyl-(1→4)-β-D-glucose)

B

Polysaccharides (oligosaccharides)

Polysaccharides are polymers of monosaccharide units. Where the number of monosaccharide units is small, the term *oligosaccharide* is used, rather than polysaccharide.

Amylose

(-α-D-glucosyl-(1 → 4)-α-D-glucosyl-)

Cellulose

The repeating unit (-β-D-glucosyl-(1→4)-α-D-glucosyl-) of *cellulose* leads to an insoluble product which is stabilized by intra- and interchain hydrogen bonds. *Cellulose* is an important constituent of plant cell walls. Having β-(1→4) bonds it cannot be digested in the human intestine, but is degraded by bacteria in the stomach of ruminants.

Starch consists of two polysaccharides, *amylose* and *amylopectin*. Amylose is a straight-chain structure containing glucose linked only by α-(1→4) bonds, whereas amylopectin contains both α-(1→4) and α-(1→6) bonds, leading to a branched structure of the type also found in glycogen in animals (see p. 197B). Starch is an important energy-storage polysaccharide of plants.

A

Glycosaminoglycans are polysaccharides important in connective tissue, in which they are attached to proteins. The glycosaminoglycan–protein complexes are referred to as *proteoglycans*.

The glycosaminoglycans consist of repeating disaccharide units, all of which bear negatively-charged groups. In some cases these negative charges are provided by sulphate groups. The repeating units of some of the most important glycosaminoglycans are shown below. Heparin is an anticoagulant. It interacts with the protein antiprothrombin and the complex inhibits enzymes of the blood-clotting process (see p. 244C).

COO^- $CH_2OSO_3^-$

Chondroitin 6-sulphate

O_3SOH_2C CH_2OH

COO^-

Dermatan sulphate

CH_2OH $CH_2OSO_3^-$

Keratan sulphate

COO^- CH_2OH

Hyaluronate

COO^- $CH_2OSO_3^-$

Heparin

B

Role of glycosaminoglycans

The synovial fluid of joints, the basement membrane and the glycocalyx that surrounds cells are all rich in glycosaminoglycans.

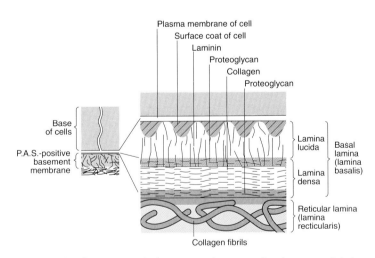

Plasma membrane of cell
Surface coat of cell
Laminin
Proteoglycan
Collagen
Proteoglycan

Base of cells

P.A.S.-positive basement membrane

Lamina lucida
Lamina densa
Basal lamina (lamina basalis)

Reticular lamina (lamina recticularis)

Collagen fibrils

Between cells in tissues, non-cellular macromolecules form a matrix known as the ground substance, rich in proteoglycans and structural glycoproteins. Associated with many types of cell, there is a more defined area known as the basement membrane, containing proteoglycans, glycoproteins (e.g. laminin) and collagen. Although generally assumed to have a structural role, it is now clear that the components of the basement membrane have more complex roles in determining metabolic and behavioural patterns of the cells contacting them, which are of fundamental importance in development, and in maintaining and regenerating tissues. Periodic acid–Schiff reagent (PAS) stains carbohydrate (periodic acid breaks the bond between adjacent hydroxy groups to form aldehydes which give a purple colour with Schiff reagent).

A

Proteoglycans have been visualized by electron microscopy.

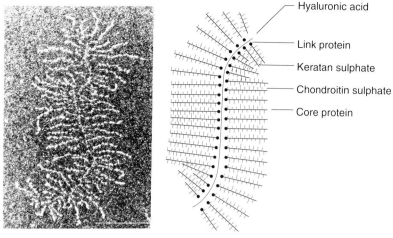

The scale in the photomicrograph represents 1 μm.

B

Glucuronic and gluconic acids

Two acids can arise from glucose. Oxidation at C-6 gives rise to *glucuronic acid*, whilst oxidation at C-1 yields *gluconic acid*.

Glucuronic acid   Glucose   Gluconic acid

Glucuronic acid is involved in the formation of glucuronides (see p. 149).

Gluconic acid is found in a phosphorylated form in the pentose phosphate pathway (6-phosphogluconic acid).

C

Uridine diphosphate glucose (UDPG)

UDPG is formed by an enzyme reaction between UTP and glucose 1-phosphate. UDPG is important in many reactions involving glucose, such as glycogen synthesis.

Glucose

Uridine diphosphate

UDPG is an example of the activation of simple molecules by addition to a nucleotide.

$$\text{UTP} + \text{Glucose 1-phosphate} \xrightarrow{\text{UDP-glucose pyrophosphorylase}} \text{UDPG} + \text{PP}_i$$

A

Nucleotide-linked sugars include *UDP-galactose* and *GDP-mannose*.

$$\text{UTP} + \text{galactose 1-phosphate} \rightarrow \text{UDP-galactose} + \text{PP}_i$$

$$\text{GTP} + \text{mannose 1-phosphate} \rightarrow \text{GDP-mannose} + \text{PP}_i$$

B

Hexose interconversions

Galactose metabolism

Galactose is phosphorylated by a specific kinase.

C

Reversible interconversion between UDP-glucose and UDP-galactose is important both in galactose degradation and in formation of UDP-galactose for synthesis of complex carbohydrates.

UDP-galactose can also be formed by the enzyme galactose 1-phosphate uridyltransferase. In this reaction, galactose 1-phosphate reacts with UDP-glucose, and glucose 1-phosphate reacts with UDP-galactose, to give the reversible cycles shown below:

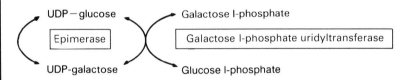

UDP-galactose is also formed by the reaction shown in A above. The enzyme responsible for this conversion develops only in later life, being absent in infants.

In galactosaemia, an inborn error of metabolism, galactokinase may be deficient, giving a mild form of the disease. Galactose 1-phosphate uridyltransferase deficiency also occurs. In the absence of UDP-galactose pyrophosphorylase during infancy, there is thus no route for the conversion of galactose 1-phosphate to UDP-galactose. This gives a more severe form of the disease.

A

The formation of UDP-glucuronic acid.

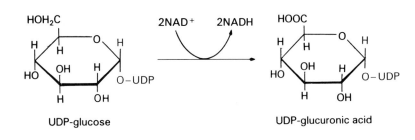

UDP-glucose   UDP-glucuronic acid

B

Mannose metabolism
Mannose can be phosphorylated by a kinase to yield mannose 6-phosphate, which can also be formed directly from, or converted to, fructose 6-phosphate.

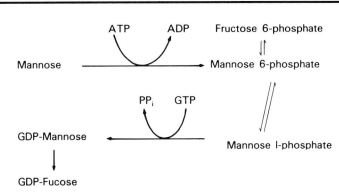

GDP-mannose and GDP-fucose are intermediates in the synthesis of complex carbohydrates. Mannose 1-phosphate acts as a precursor for the synthesis of GDP-mannose, from which GDP-fucose can be formed.

C

Mannose derivatives participate in the formation of *N*-acetylneuraminic acid (sialic acid) but the route of formation is from glucosamine 6-phosphate, formed by reaction of glutamine with the carbonyl group of fructose 6-phosphate.

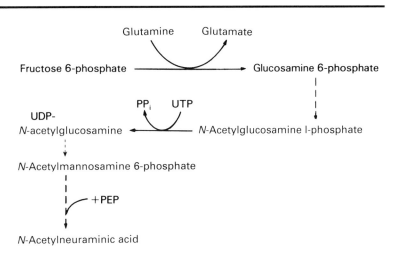

A

The role of sugars in detoxification

The introduction of a hydroxyl group into an aromatic ring in itself increases the water solubility of the compound. There are, in the smooth endoplasmic reticulum, enzymes that catalyse addition of a single oxygen atom (i.e. mono-oxygenases). This oxidation involves a cytochrome, called cytochrome P450, because it has an absorption maximum at 450 nm. The overall reaction is

$$RH + O_2 + 2e^- + 2H^+ \rightarrow ROH + H_2O$$

The hydrogens needed for the reaction are derived from NADPH, and are transferred by a flavoprotein named cytochrome P450 reductase.

These enzymes are increased 25-fold in liver by phenobarbital, which induces massive proliferation of the endoplasmic reticulum of liver.

Compounds possessing a hydroxyl group (including those formed by the reaction described above) are often converted to glucuronides (glycosides of glucuronic acid) by reaction with UDP-glucuronic acid.

B

Formation of glucuronides.

The excretion of compounds that are not very water-soluble presents difficulties that can be overcome by metabolic conversions which render the material more soluble in water.

Thus, after the introduction of a hydroxyl group into an aromatic ring (see above), it can be conjugated with glucuronic acid, which also reduces the toxicity of the compound.

UDP-glucuronic acid

Phenol

Phenyl glucuronide

a glycoside of glucuronic acid

glucuron · · · ide

α-Glucuronidases and β-glucuronidases hydrolyse the glycosidic bond of glucuronides and, as their names imply, are specific for the configuration on C-1 of glucuronic acid.

C

Another route of detoxification involves the formation of glycine conjugates.

Benzoic acid

Glycine

Hippuric acid

Many steroid metabolites are excreted as glucuronides. Other compounds, especially acids, may be conjugated with glycine, and excreted as glycine conjugates.

NITROGEN METABOLISM

Body protein is in a constant state of turnover through the activity of synthetic and degradative pathways. Activity of the *urea cycle*, a pathway for the removal of the amino group as a prelude to the breakdown of the residual carbon chain, leads to the excretion of the amino nitrogen as *urea*. If the excretion of urea nitrogen exceeds the intake of nitrogen in food, such as in starvation, a state of *negative nitrogen balance* pertains; if the reverse is the case, as in a young growing animal, the animal is in *positive nitrogen balance*. A perfect balance between the two processes leads to *nitrogen equilibrium*. The first step in the breakdown of an amino acid involves removal of the amino group by one or other of the *transaminases*. A pathway exists for the breakdown of the carbon skeleton of every amino acid.

Some amino acids can be synthesized in mammals by transfer of an amino group *(transamination)* to a readily available *keto acid (oxo acid)* such as oxaloacetate (for aspartate) or oxoglutarate (for glutamate). Acids that can be made in this way are *non-essential*, those that must be provided in the diet being *essential amino acids*.

Reaction of creatine phosphate with ADP catalysed by *creatine kinase* provides an immediate source of ADP in muscle. The daily urinary excretion of creatinine, a cyclized products of creatine, is sometimes used as an index of muscular activity.

Amino acids act as precursors for compounds such as thyroxine, histamine and adrenaline. The synthesis of haem is an important specialized pathway.

A

Essential and non-essential amino acids

The amino acids that are used for protein synthesis cannot all be synthesized in the body. Those that can be synthesized are termed *non-essential amino acids*, whilst those that cannot be synthesized are termed *essential amino acids*.

Essential amino acids
Histidine
Isoleucine
Leucine
Valine
Lysine
Methionine
Threonine
Tryptophan
Phenylalanine

Non-essential amino acids synthesized directly by transamination from metabolites readily available from major pathways

Amino acid	Metabolite from which synthesized
Alanine	Pyruvate
Aspartic acid	Oxaloacetate
Glutamic acid	Oxoglutarate

Non-essential amino acids synthesized by special pathways.

Amino acid	Metabolite from which synthesized
Proline	from glutamate via ornithine
Glycine	Pyruvate
Serine	Glycine
Cysteine	Methionine
Tyrosine	Phenylalanine
Arginine*	from glutamate via ornithine

*Arginine may be needed in the diet for adequate growth rates in the young.

If it is to provide adequate nutrition (i.e. to have a high 'biological value'), protein must contain a balanced proportion of all of the essential amino acids. Animal proteins (meat, eggs) are a good source, but a number of plant proteins may be deficient in one or more of the essential amino acids. Corn (maize) is deficient in lysine, and rice protein is deficient in lysine and threonine. New strains are being bred with more adequate proportions of these amino acids.

B

Transamination and oxidative deamination

Amino groups can be removed by oxidative deamination or by transamination, and can be added by transamination.

The general reaction for a transamination reaction is

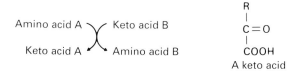

Enzymes of this type are termed transaminases, or, in more recent terminology, *aminotransferases*.

Individual keto acids are now designated as the oxo derivative, e.g. oxoglutarate. The term keto acid is still used to designate the class of compound.

A

Reactions that remove the amino group, or add it to a chain that previously lacked it, are central to amino acid metabolism.

Glutamate dehydrogenase is one of the most important enzymes removing the amino group.

$$
\begin{array}{c}
COO^- \\
| \\
CH_2 \\
| \\
CH_2 \\
| \\
CH-NH_3^+ \\
| \\
COO^-
\end{array}
\; + NAD^+ + H_2O \rightleftharpoons \;
\begin{array}{c}
COO^- \\
| \\
CH_2 \\
| \\
CH_2 \\
| \\
C=O \\
| \\
COO^-
\end{array}
\; + NADH + H^+ + NH_4^+
$$

Glutamate   2-Oxoglutarate*

Glutamate dehydrogenase reaction

* α-Ketoglutarate, the original name for 2-oxoglutarate, is still frequently used.

B

A transaminase reaction of central importance, involving alanine transaminase, transfers the amino group of alanine to oxoglutarate to form glutamate.

$$
\begin{array}{c}
CH_3 \\
| \\
CH-NH_3^+ \\
| \\
COO^-
\end{array}
\; + \;
\begin{array}{c}
COO^- \\
| \\
CH_2 \\
| \\
CH_2 \\
| \\
C=O \\
| \\
COO^-
\end{array}
\; \rightleftharpoons \;
\begin{array}{c}
CH_3 \\
| \\
C=O \\
| \\
COO^-
\end{array}
\; + \;
\begin{array}{c}
COO^- \\
| \\
CH_2 \\
| \\
CH_2 \\
| \\
CH-NH_3^+ \\
| \\
COO^-
\end{array}
$$

Alanine   Oxoglutarate   Pyruvate   Glutamate
Alanine transaminase (ALT)
(or glutamate–pyruvate transaminase (GPT))
reaction

C

The most important of the enzymes that carry out oxidative deamination is glutamate dehydrogenase. Operating in conjunction with transaminases, it can convert the amino group of most amino acids to free ammonia.

Amino acid ⤸ Oxoglutarate ⤸ NADH + H⁺ + NH₄⁺
Keto acid ⤹ Glutamate ⤹ NAD⁺ + H₂O

Many transaminases   Glutamate dehydrogenase

D

Alanine transaminase can act as a link between glutamate dehydrogenase and amino acids that do not react directly with oxoglutarate.

Amino acid ⤸ Pyruvate ⤸ Glutamate ⤸ NAD⁺ + H₂O
Keto acid ⤹ Alanine ⤹ Oxoglutarate ⤹ NADH + H⁺ + NH₄⁺

Many transaminases utilize alanine as substrate   Alanine transaminase   Glutamate dehydrogenase

A

Transaminases that transfer amino groups to oxaloacetate to form aspartate are important as this is a route for incorporating the $-NH_2$ group into urea.

Glutamate ⇌ Oxaloacetate

Oxoglutarate ⇌ Aspartate

Aspartate transaminase (AST)
(or glutamate-oxaloacetate transaminase (GOT))
reaction

B

Another source of free ammonia is the monoamine oxidase reaction.

Vanillylmandelate in the urine can be measured as an index of catecholamine metabolism.

Biological amines (e.g. adrenaline) $\xrightarrow{\text{Monoamine oxidase (MAO)}}$ NH_4^+ + Aldehydes

For example:

Noradrenaline $\xrightarrow{\text{MAO}}$

Methylation and oxidation

Vanillylmandelate

C

Cooperation between the various transaminase and deaminase enzymes leads to the formation of free ammonium ions and net formation of aspartate. The ammonium ion and the amino group of the aspartate provide the two nitrogens of urea and are thus eliminated from the body.

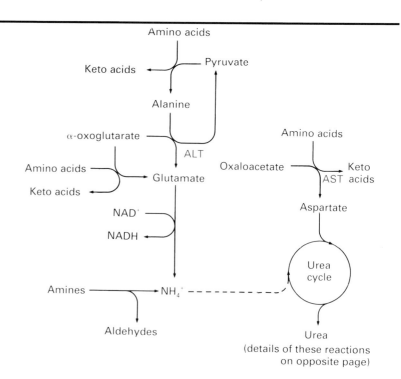

Urea
(details of these reactions on opposite page)

The urea cycle

The formation of urea takes place through a complex series of reactions known as the urea cycle. The start of the urea cycle may be regarded as the reaction between the amino acid ornithine and carbamoyl phosphate, to form citrulline. Ornithine is finally regenerated in the reaction that forms urea, thus giving rise to the cyclic form of the overall reaction.

Free ammonia is needed for the synthesis of carbamoyl phosphate. This reaction, and the formation of glutamine, are important in maintaining a low concentration of free ammonia, which is toxic.

The synthesis of glutamine (from glutamate, NH_4^+ and ATP) by glutamine synthase also removes free NH_4^+. Asparagine is formed from glutamine and aspartate (amide transfer) by asparagine synthase. Some tumour cells lack asparagine synthase. Certain forms of leukaemia can be treated by administering asparaginase, an enzyme that hydrolyses the amide group of asparagine to convert it back to asparate, thus lowering blood levels of asparagine and starving the tumour cells.

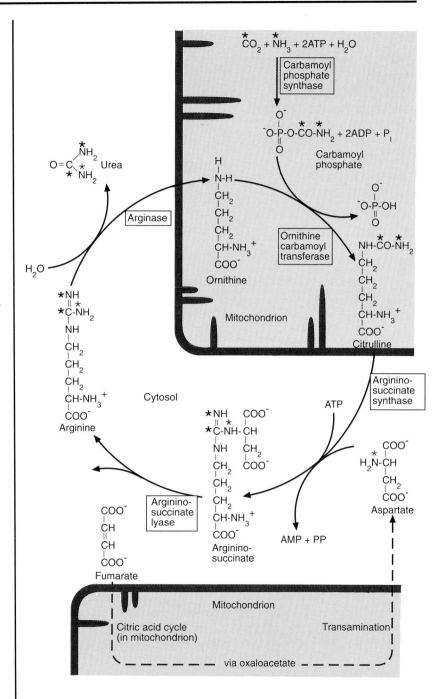

The importance of aspartate in providing one of the nitrogens of urea is noteworthy. This nitrogen is derived from transamination reactions that culminate in the aspartate transaminase reaction.

A

Inherited defects of urea cycle enzymes.

Deficiency of urea cycle enzymes leads to metabolic disorders, some of which are listed in the table.

Enzyme deficient	No. of cases reported	Type of inheritance	Characteristic features
Carbamoyl phosphate synthase	26	Autosomal recessive	Hyperammonaemia and aminoacidaemia without orotic aciduria (see p. 96)
Ornithine carbamoyltransferase	110	X-linked dominant	Hyperammonaemia and aminoacidaemia with orotic aciduria
Argininosuccinate synthase	53	Autosomal recessive	Citrullinaemia and citrullinuria
Argininosuccinate lyase	60	Autosomal recessive	Mild argininosuccinic acidaemia with argininosuccinic aciduria
Arginase	13	Autosomal recessive	Arginaemia (variable with dietary load) and argininuria

The number of cases is, of course, approximate, and increases with time, but gives an indication of relative incidence. The prognosis varies with the severity of the disease. For example, a neonatal form of carbamoyl phosphate synthase deficiency is associated with complete or almost complete absence of the enzyme and normally results in death in the neonatal period. However, partial deficiency of the enzyme also occurs, with correspondingly longer survival to early adulthood or later.

B

An important regulatory mechanism involves the compound *N-acetylglutamate*. The synthesis of this is regulated by the concentration of arginine.

Control of the urea cycle

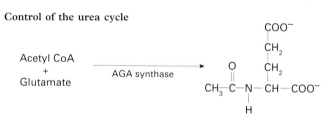

N-Acetylglutamate (AGA)

AGA synthase activity is stimulated by high protein intake.

C

Carbamoyl phosphate synthase has an absolute requirement for *N*-acetylglutamate. Thus, indirectly it is controlled by the level of arginine, which stimulates AGA synthase.

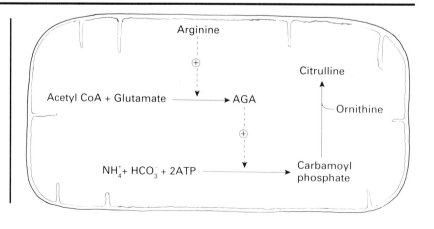

A

The synthesis of non-essential amino acids

Synthesis of alanine, aspartate and glutamate

These three amino acids are synthesized by direct transamination from pyruvate, oxaloacetate and 2-oxoglutarate reacting with another amino acid.

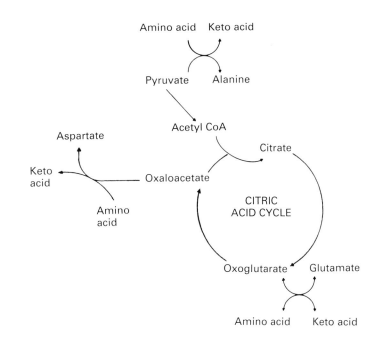

B

Synthesis of glycine and serine

Both serine and glycine are synthesized from 3-phosphoglyceric acid, derived from glucose as an intermediate in the glycolytic pathway.

$$\begin{array}{ccc}
CH_2O\text{(P)} & & CH_2O\text{(P)} \\
| & \xrightarrow[\text{NAD}^+ \, \text{NADH}]{\text{3-Phosphoglyceric-acid dehydrogenase}} & | \\
CHOH & & CO \\
| & & | \\
COO^- & & COO^-
\end{array}$$

3-Phosphoglyceric acid

3-Phosphohydroxypyruvic acid

Transamination

$$\begin{array}{ccc}
CH_2O\text{(P)} & & CH_2OH \\
| & \xrightarrow[\quad P \quad]{\text{Phosphoserine phosphatase}} & | \\
CH-NH_3^+ & & CH-NH_3^+ \longrightarrow \text{Glycine} \\
| & & | \qquad \text{(Loss} \\
COO^- & & COO^- \quad \text{of formaldehyde)}
\end{array}$$

3-Phosphoserine

Serine

C

Glycine and serine are interconvertible in a reversible reaction involving a folate coenzyme.

$$\begin{array}{ccc}
CH_2NH_3^+ & & CH_2OH \\
| & & | \\
COO^- + \text{5,10-Methylene } H_4\text{-folate} & \rightleftharpoons & CH-NH_3^+ + H_4\text{-folate} \\
& & | \\
\text{Glycine} & & COO^- \\
& & \text{Serine}
\end{array}$$

A

The synthesis of cysteine

The synthesis of cysteine from methionine involves *S-adenosylmethionine* (see below).

S-Adenosylmethionine is an important intermediate in the transfer of a methyl group from methionine to a variety of acceptor molecules.

$$CH_2-S-CH_3$$
$$|$$
$$CH_2$$
$$|$$
$$CH-NH_3^+$$
$$|$$
$$COO^-$$

$+ ATP \longrightarrow$

$$CH_3$$
$$|$$
$$CH_2-S-CH_2$$
$$| \quad +$$
$$CH_2$$
$$|$$
$$CH-NH_3^+$$
$$|$$
$$COO^-$$

adenine

$+P_i +PP_i$

Methionine

S-Adenosylmethionine ('active' methionine)

B

S-Adenosylmethionine is converted to homocysteine in a reaction in which its methyl group is transferred to *guanidoacetic acid*. Homocysteine can then form cysteine after reaction with serine.

S-Adenosylmethionine

$NH_2-C=NH_2^+$
$NH-CH_2COOH$ Guanidoacetic acid

S-Adenosylhomocysteine

$NH_2-C=NH_2^+$
H_3CN-CH_2COOH Creatine

Enzyme in liver

Adenosine $+ CH_2SH$
$$|$$
$$CH_2$$
$$|$$
$$CH-NH_3^+$$
$$|$$
$$COO^-$$

Homocysteine

Serine

$$CH_2-S-CH_2$$
$$| \qquad |$$
$$CH_2 \qquad CN-NH_3^+$$
$$| \qquad |$$
$$CH-NH_3^+ \quad COO^-$$
$$|$$
$$COO^-$$

Cystathionine

$+H_2O$

$$CH_2OH \qquad HS-CH_2$$
$$| \qquad\qquad |$$
$$CH_2 \qquad + \quad CH-NH_3^+$$
$$| \qquad\qquad |$$
$$CH-NH_3^+ \qquad COO^-$$
$$|$$
$$COO^-$$

Homoserine   Cysteine

C

The synthesis of proline from ornithine

Ornithine is synthesized from arginine by the action of arginase. Proline can then be synthesized from ornithine.

$$CH_2NH_3^+ \quad NH_4^+ \quad HC=O$$
$$| \qquad\qquad\quad |$$
$$CH_2 \qquad\qquad CH_2 \quad \text{Cyclizes}$$
$$| \qquad\qquad\quad |$$
$$CH_2 \quad H_2O \quad CH_2$$
$$| \qquad\qquad\quad |$$
$$H_3^+N-CH-COO^- \quad H_3^+N-CH-COO^-$$

Ornithine

Glutamic semi-aldehyde

NADH NAD$^+$

$$H_2C-CH_2$$
$$| \qquad |$$
$$HC \quad HC-COOH$$
$$\diagdown \! / $$
$$N$$

Δ^1-Pyrroline-5-carboxylate

$$H_2C-CH_2$$
$$| \qquad |$$
$$H_2C \quad HC-COO^-$$
$$\diagdown \! / $$
$$N$$
$$|$$
$$H$$

Proline

The catabolism of phenylalanine and tyrosine

In general, the catabolic pathways that degrade the essential amino acids are complex and specialized. However, the pathways involved in the catabolism of phenylalanine and tyrosine have a special interest because deficiencies of enzymes in these pathways are associated with hereditary diseases of historical importance. *Alcaptonuria* and *albinism* were two of the first diseases to be ascribed to a genetic origin, and *phenylketonuria* was the first genetic disease for which a successful therapy was devised.

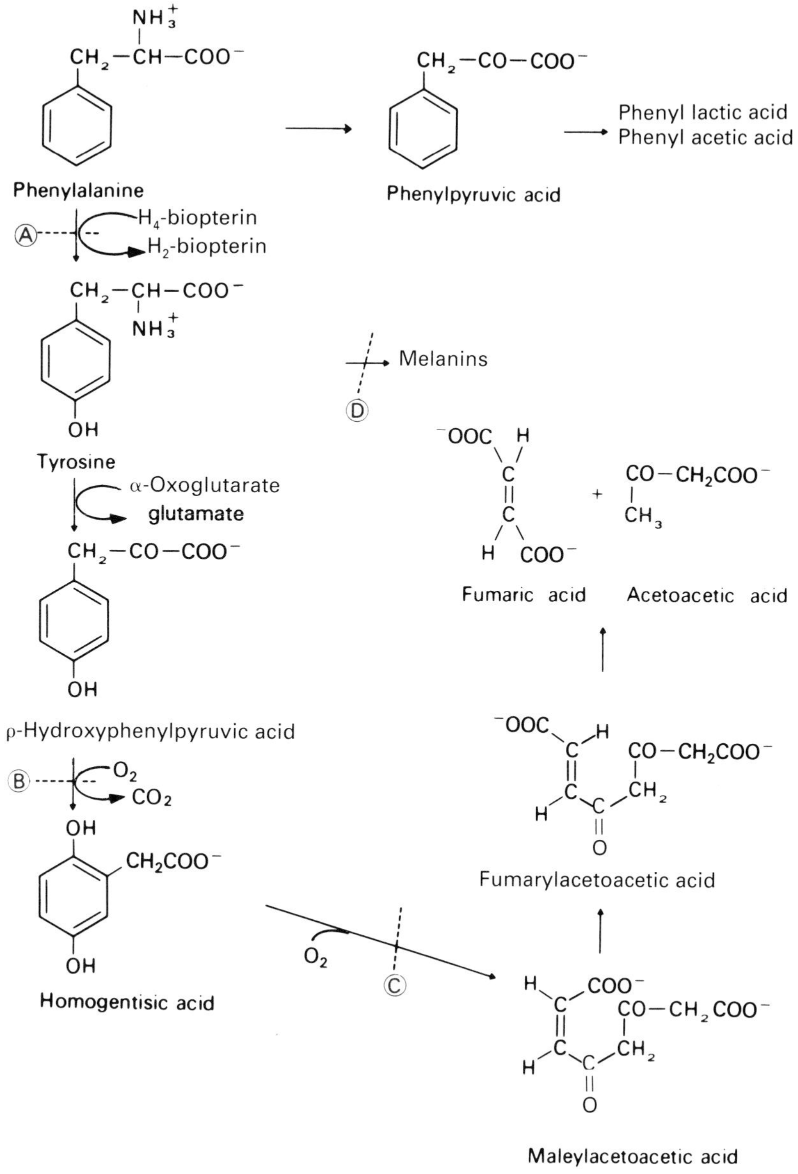

Phenylalanine

Tyrosine

p-Hydroxyphenylpyruvic acid

Homogentisic acid

Phenylpyruvic acid

Phenyl lactic acid
Phenyl acetic acid

Melanins

Fumaric acid   Acetoacetic acid

Fumarylacetoacetic acid

Maleylacetoacetic acid

Enzyme A missing—*phenylketonuria* produced. Blockage of the main pathway diverts phenylalanine to phenylpyruvic acid and derivatives whose accumulation leads to mental deficiency (see p. 160B).
Enzyme B missing—*tyrosinosis* produced. A very rare condition in which p-hydroxyphenylpyruvic acid and tyrosine are excreted in the urine.
Enzyme C missing—*alcaptonuria* produced. Homogentisic acid is excreted in the urine which may turn black on standing.
Enzyme D missing—*albinism* produced. The natural melanin pigments of skin, hair and eyes are not formed.

A

The hydroxylation of phenylalanine to tyrosine is carried out by an enzyme that utilizes molecular oxygen for the introduction of an oxygen atom into the aromatic ring. This enzyme is a *monooxygenase* or *mixed-function oxidase*.

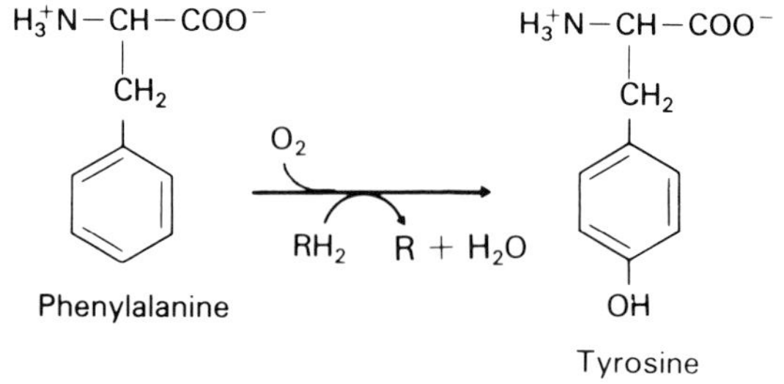

Phenylalanine → Tyrosine

Tyrosine hydroxylase and tryptophan hydroxylase have essentially similar mechanisms to phenylalanine hydroxylase.

B

The phenylalanine hydroxylase reaction.

In the phenylalanine hydroxylase reaction, RH_2 is 5,6,7,8-tetrahydrobiopterin, which is oxidized to 6,7-dihydrobiopterin. This can then be reduced by dihydrofolate reductase to regenerate the reduced form.

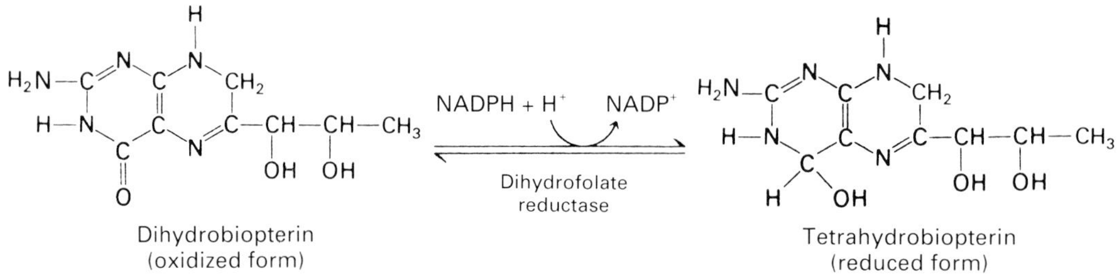

Dihydrobiopterin (oxidized form) → Tetrahydrobiopterin (reduced form)

C

Phenylketonuria.

Phenylketonuria was the first inborn error to receive rational therapy as a result of the discovery of the molecular defect. Elimination of most of the phenylalanine from the diet leads to amelioration of symptoms.

Congenital absence of phenylalanine hydroxylase occurs in phenylketonuria. In this condition the inability to convert phenylalanine to tyrosine results in metabolism of large amounts of phenylalanine by the transaminase reaction to form phenylpyruvate (the 'ketone' found in the urine).

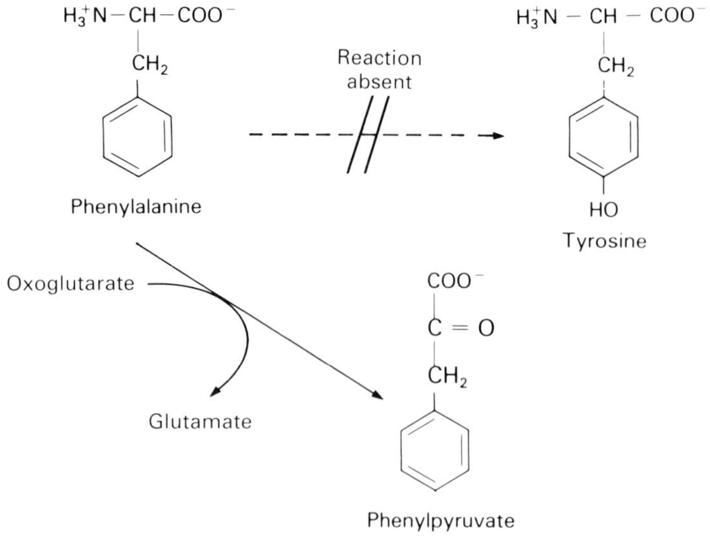

Phenylalanine   Reaction absent   Tyrosine

Oxoglutarate → Glutamate

Phenylpyruvate

A

The formation of thyroid hormones

The synthesis of the hormones thyroxine (T_4) and triiodothyronine (T_3) involves the iodination of tyrosine, whilst bound as tyrosine residues of the protein *thyroglobulin*.

L-Tyrosine

L-3-Monoiodotyrosine

L-3,5-Diiodotyrosine

L-3,5,3'-Triiodothyronine (triiodothyronine)

2 molecules

L-3,5,3'5'-Tetraiodothyronine (thyroxine)

B

Thyroglobulin, the protein containing the tyrosine residues from which thyroxine is formed is contained in the follicle within the thyroid gland. Colloid containing thyroglobulin is taken into vesicles in which it is degraded. Proteolysis releases T_4 and T_3, which are then extruded into the blood capillaries.

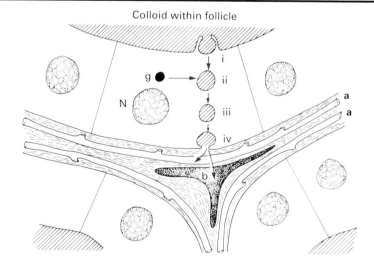

Diagrammatic representation of the breakdown of thyroglobulin in the thyroid gland. (i) Phagocytosis of colloid to form PAS*-positive vesicle. (ii) Transference of hydrolytic enzymes from dense granule (*g*). (iii) Digestion of colloid with progressive loss of PAS-positive material. (iv) Extrusion of iodothyronines mainly to the thyroid blood capillary (*a*). N, nucleus; b, lymphatic capillary.

* PAS, refers to the periodic acid–Schiff reaction, specific for carbohydrate.

A

The formation of adrenaline and noradrenaline

These amines belong to a group of compounds known as the *catecholamines*.

Adrenaline is synthesized from tyrosine in the adrenal medulla.

B

Catecholamines are important in the nervous system.

Parkinson's disease is caused by loss or inhibition of dopamine activity.

Noradrenaline is a neurotransmitter in the central nervous system and at sympathetic nervous system nerve endings. Dopamine also is involved in neurotransmission, and is particularly important in the *substantia nigra*, a part of the midbrain so-called because of its content of *melanin*, a black pigment of which dopaquinone is a precursor, formed from tyrosine by *tyrosinase*. The substantia nigra is important in mediating motor activity of the brain. Destruction of the nigrostriatal cells in the substantia nigra causes *Parkinson's disease*. Symptoms of the disease, such as tremor, can be alleviated by administration of L-dopa, which crosses the blood–brain barrier and is converted to dopamine (which does not cross the blood–brain barrier and therefore cannot be administered directly).

Related in structure to dopamine are mescaline, a hallucinogenic drug extracted from the *mescal buttons* of the spineless peyote cactus, and amphetamine, a drug that produces a feeling of mental alertness and well-being.

A

The metabolism of histidine

Formation of histamine.

The important *autacoid*★ histamine is released by *mast cells* in respone to a variety of stimuli, including IgE (mediating allergic responses).

Histamine is produced as a result of the action of *histidine decarboxylase*.

★ The term 'autacoid' denotes any compound secreted into the blood by one organ to act on another.

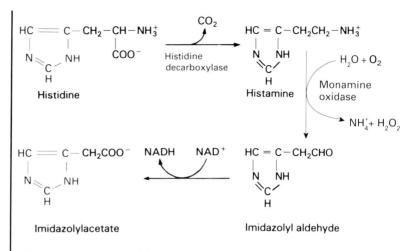

Histamine is destroyed by monoamine oxidase, forming imidazolyl aldehyde, which is oxidized to imidazolylacetate, in which form it is excreted in the urine (as a conjugate with ribose 5'-phosphate).

B

Degradation of histidine.

The main degradative pathway for histidine yields glutamic acid.

Formiminoglutamic acid (FIGLU) may be excreted in quantity in the urine in certain pathological conditions such as formiminotransferase deficiency. As a folate coenzyme is required for its metabolism, it accumulates and is excreted in the urine in folate deficiency.

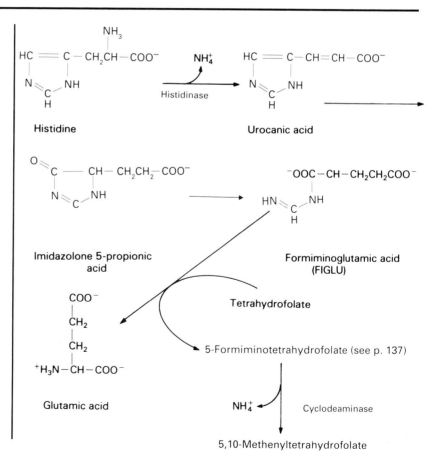

The metabolism of tryptophan

Tryptophan serves as the precursor to two important compounds, *serotonin* and *nicotinamide*. Serotonin a (5-hydroxytryptamine), neurotransmitter, is formed after hydroxylation and decarboxylation of tryptophan. The synthesis of the nicotinamide moiety of the nicotinamide nucleotides is initiated by the enzyme *tryptophan dioxygenase (tryptophan pyrrolase)*, which opens the indole ring.

Patients with *carcinoid*, a tumor of the *argentaffin cells* of the pancreas, secrete large amounts of serotonin.

Indole

Tryptophan is indolylalanine

In B_6 deficiency, metabolism of 3-hydroxykynurenine is depressed, due to the fact that the enzyme is B_6-dependent. Under these conditions, much more tryptophan is metabolized to xanthurenic acid, an event that can be utilized to assess the adequacy of B_6 nutritional status.

A

Glutathione and oxygen toxicity

The tripeptide glutathione is found in all mammalian cells, except the neurone, and in many other non-mammalian cell types.

```
COO⁻
|
CH—NH₃⁺
|
CH₂
|
CH₂        CH₂—SH
|          |
CO—NH—CH—CO—NH—CH₂—COO⁻
```

Glutathione (γ-glutamyl-cysteinyl-glycine)

Glutathione is characterized by its unique γ-peptide bond which is not attacked by peptidases, but is a substrate for γ-glutamyltransferase (see p. 77A). The sulphydryl is the functional group primarily responsible for the properties of glutathione (often abbreviated to GSH). One of the most important reactions in which glutathione participates, catalysed by glutathione peroxidase, inactivates hydrogen peroxide and other peroxides.

$$GSH + H_2O_2 \text{ (or } RO_2H) \rightarrow GSSG + 2H_2O \text{ (or } ROH + H_2O)$$

GSSG is then reduced back to GSH by reaction with NADPH under the action of glutathione reductase.

B

Free radicals

A covalent chemical bond is formed by two electrons that share a bonding orbital. If a bond is broken in such a way that each of the resulting molecules takes one of the pair of electrons, each product is a *free radical*, in that it has an unpaired electron. Thus, loss of a hydrogen atom from an alkyl chain forms a free radical. Contact with other molecules, which have a complete number of bonding electrons, can lead to transfer of the lone free radical electron to form another free radical, and on repetition of this a chain reaction will ensue. This can be stopped by a molecule that can absorb a free electron to form a stable compound. Such molecules include α-tocopherol (vitamin E, see. p. 243B).

C

Oxygen toxicity

Although oxygen is essential to almost all forms of life, it can give rise to highly toxic molecular species, especially when present at higher concentrations than in air at normal atmospheric pressure (referred to as *hyperbaric oxygen*).

Electron configuration of ground state oxygen and the superoxide ion

Orbital	σ	σ⋆	σ	σ⋆	σ	π	π⋆
Subshell	1s	1s	2s	2s	2p	2p	2p

Ground state oxygen

Superoxide ion (O_2^-)

One of the most important of the reactive molecular species is the superoxide ion, a free radical. *Singlet oxygen* also has higher reactivity than oxygen. Two forms of singlet oxygen occur, one in which both of the π⋆ 2p orbitals are occupied, each containing an electron but of opposite spins, and one in which only one of these orbitals is occupied, with two electrons each of opposite spin. *Superoxide dismutase* catalyses the reaction of two molecules of the superoxide ion with two protons to form one molecule of O_2 and one of H_2O_2. Arrows indicate electrons and direction of spin.

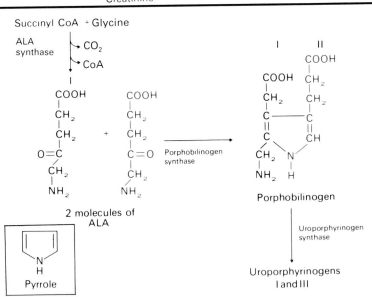

A

Phosphocreatine

In muscle, during contraction, ATP is for the most part derived from glycolysis, for which muscle glycogen acts as initial substrate. However, before ATP from glycolysis becomes available, *phosphocreatine* acts as a temporary source of ATP.

Phosphocreatine, and creatine, may break down to creatinine. This breakdown is fairly constant over a 24 hour period. Urinary creatinine is sometimes used as an index of urinary excretion.

B

The synthesis of haem

The synthesis begins with the compound δ-*aminolaevulinic acid* (ALA), which itself is synthesized by a condensation between glycine and succinyl CoA, with loss of CoA and CO_2, catalysed by *ALA synthase*.

Two molecules of ALA condense to form compounds that are converted to *porphobilinogen*, four molecules of which then polymerize to the tetrapyrroles *uroporphyrinogens I and III*. Only *uroporphyrinogen III* is used for haem synthesis.

Both ALA synthase and ALA dehydratase are inhibited by haem. In acute intermittent porphyria (see p. 168A), haem synthesis is decreased, and the feedback inhibition on ALA synthase is decreased. Certain drugs such as barbiturates and oestrogens induce this enzyme and may precipitate an attack.

Synthesis of ALA occurs in the mitochondrion whilst porphobilinogen synthesis is cytoplasmic.

Uroporphyrinogens I and II are identical except that the order of the carboxymethyl and carboxyethyl side-chains on one pyrrole is reversed.

Uroporphyrinogen I is synthesized in smaller amounts and converted to several products essentially similar to those synthesized from uroporphyrinogen III. However, the I-series do not form haem and are excreted in the urine.

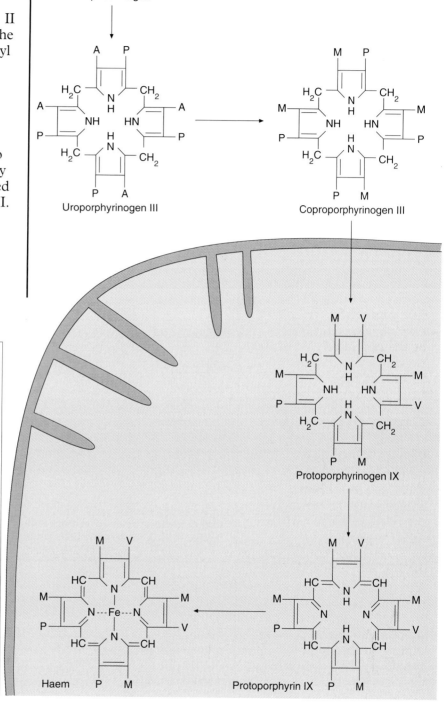

Metabolism of uroporphyrinogen III to haem involves the side chains of the pyrroles and the hydrogen atoms on the pyrrole nitrogens.

The synthesis of coproporphyrinogen III from porphobilinogen is cytoplasmic. The remaining steps in the synthesis of haem occur in the mitochondrion.

A

The *porphyrias* are inherited diseases of porphyrin metabolism

Name	Cell affected	Type of inheritance	Accumulating materials	Diagnosis
Erythropoietic uroporphyria	Erythrocyte	Autosomal recessive	Uroporphyrin I Coproporphyrin	1. Both compounds in urine and faeces and red cells 2. Red cells fluoresce under ultraviolet light
Erythropoietic protoporphyria	Erythrocyte	Autosomal dominant	Protoporphyrin	Excretion of protoporphyrins in faeces
Acute intermittent porphyria	Liver	Autosomal dominant	ALA Porphobilinogen	1. Both compounds in urine 2. Urine turns to deep red in sunlight as porphobilinogen is converted to uroporphyrin

B

Bile pigments

At the end of the life of the red cell it is broken down by cells of the reticuloendothelial system and the haem is degraded to *biliverdin*. This involves oxidation of one of the interpyrrole bridges. Reduction of biliverdin yields *bilirubin*. These compounds are found in the bile and are known as the bile pigments.

In the liver, bilirubin reacts with two molecules of UDP-glucuronate to form a bisglucuronide, which has a glucuronic acid residue in glucuronide linkage to each of the propionic residues.

C

Further metabolism in intestine and kidney yields other pigments that are found in faeces and urine.

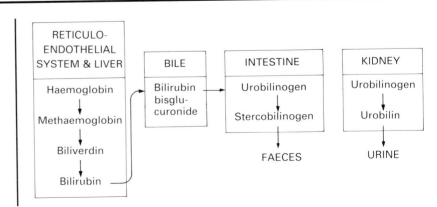

8

CARBO-HYDRATE AND FAT METABOLISM

A. Oxidative catabolism

Any sequence of chemical reactions proceeds in a direction giving a negative free energy change. In general, the molar free energy of complex molecules is high, so that their formation must be coupled with a reaction involving a large negative free energy change. In synthetic (*anabolic*) processes this can often be achieved by involving ATP as a reactant, since hydrolysis of the phosphate anhydride bond is accompanied by a substantial decrease in free energy. A continual regeneration of ATP to drive synthetic processes is achieved by coupling the formation of ATP with the oxidation of glucose (*glycolysis*) or fatty acid (*β-oxidation*), both of these *catabolic* processes being *exergonic*. The products of glycolysis and β-oxidation are further oxidized to CO_2 and H_2O by the *electron transport chain* of haem-containing cytochromes, with further formation of ATP (*oxidative phosphorylation*).

A

Forces driving metabolic processes

All life depends on the utilization of energy emitted by the sun.

This is utilized by plants to convert atmospheric CO_2 to organic substances, which are consumed by herbivorous animals. Oxidation of ingested nutrients by mammals releases energy that is used to drive chemical synthesis and other work.

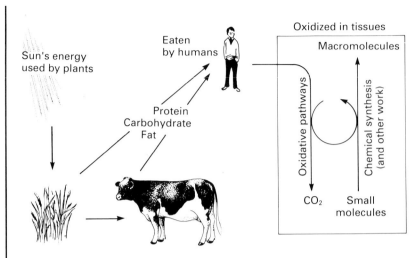

Chemical reactions tend always to proceed towards equilibrium. If a product of a chemical reaction (such as a gas) is continually removed, the reaction never reaches equilibrium and will proceed as long as reactants (substrates in an enzymic reaction) are available. Thus, continual loss of CO_2 (an end-product of mammalian oxidation pathways) through the lungs ensures that oxidation processes can act as a constant driving force for other reactions in the body. Although one of the reactants, O_2, is also a gas, it is in much greater proportion in the atmosphere than is CO_2 and is concentrated at the tissues by the action of haemoglobin.

B

Synthetic (*anabolic*) pathways depend on a supply of certain coenzymes generated by degradative (*catabolic*) pathways.

Note: NADPH is not itself generated by many major oxidative pathways, and is not exclusively used for reductive processes in synthetic pathways but, for simplicity, only the one nicotinamide coenzyme is shown.

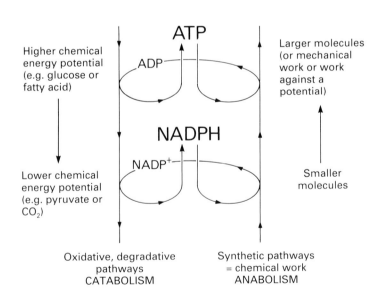

Certain coenzyme systems (ATP/ADP, NADH/NAD$^+$, NADPH/NADP$^+$) are involved in many reactions within a cell. Of these coenzymes, ATP, NADH and NADPH often participate as reactants in synthetic reactions (or, in the case of ATP, as an essential component of a system doing work, such as muscular contraction, or transport of molecules across membranes).

A

The pathways that generate the major part of the ATP and NADH formed in the cell are:

1. Glycolysis
2. Fatty acid oxidation
3. Citric acid cycle
4. Electron transport chain.

NADPH can be formed from $NADP^+$ and NADH by a mitochondrial transhydrogenase, linked to the electron transport system and essentially irreversible.

$$NADH + NADP^+ \rightarrow$$
$$NADPH + NAD^+$$

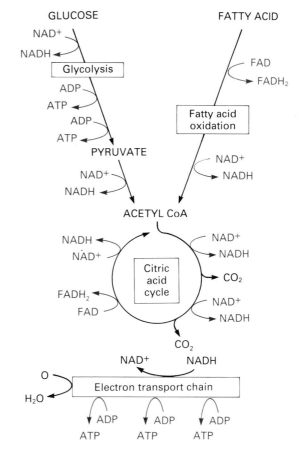

In the cytosol, transfer of hydrogen between NADH and NADP+ is brought about by malate dehydrogenase and the malic enzyme (see p. 217).

B

The carbons of glucose are already partially oxidized, whilst most of those of a fatty acid are fully reduced.

Glucose

$$CH_3CH_2CH_2CH_2CH_2CH_2CH_2CH_2CH_2CH_2CH_2CH_2CH_2CH_2CH_2COOH$$

Palmitic acid (a fatty acid)

Most of the carbons of a fatty acid are in the fully reduced state. Thus, fatty acids can potentially undergo a greater number of oxidative steps for each carbon than can carbohydrates. Thus, they are an ideal substance for use as an energy store. Glycogen (a polymer of glucose) is, on the other hand, stored as an immediate reserve of blood glucose.

A

An overall plan of
metabolic pathways.

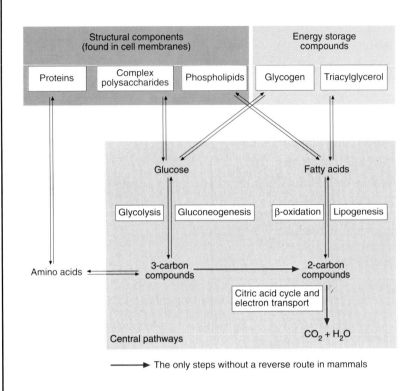

B

Because it is in a more
highly reduced state, fat has
a higher energy potential
than has carbohydrate, and
thus its oxidation can yield
more energy.

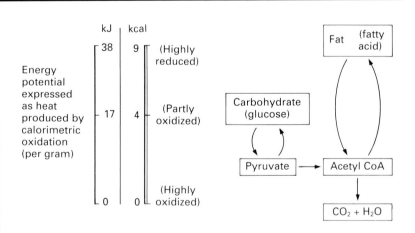

A

β-oxidation of fatty acids

Fatty acids are oxidized in the mitochondria. A fatty acid is degraded step by step by the sequential removal of two-carbon units. Coenzyme A is added to the carboxyl group of the fatty acid to form the fatty acyl CoA derivative. Each two-carbon unit is liberated from the fatty acid as acetyl CoA. The oxidation of the fatty acid alkyl chain, by enzymes that utilize FAD and NAD^+, yields a compound that is at the oxidation state of an acid (acetyl CoA is the condensation product of acetic acid and coenzyme A).

Note the similarity to a sequence of reactions in the citric acid cycle (see p. 179). The reaction sequence is sometimes termed β-oxidation because the oxygen is added at the β-carbon.

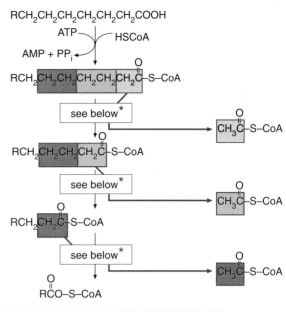

B

A summary of fatty acid oxidation.

Control of the rate of fatty acid oxidation is regulated by the rate of transport of the fatty acids into the mitochondrion. This depends on the formation of carnitine esters (see p. 203B).

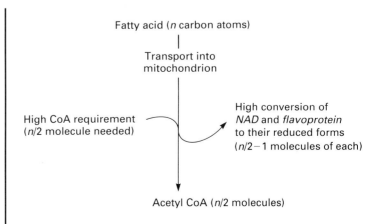

A

The glycolytic pathway

The oxidation of glucose to pyruvate (termed glycolysis)

Successive phosphorylation of C-1 and C-6 of glucose utilizes initially two molecules of ATP. These phosphorylations and an isomerization bring about the conversion of glucose to fructose 1,6-bisphosphate. After the split of the six-carbon sugar to two three-carbon molecules (by aldolase) there is immediate oxidation of the aldehyde to an acid with simultaneous formation of an anhydride bond between the acid and a phosphate group. This complex and important reaction is achieved by glyceraldehyde phosphate dehydrogenase, utilizing NAD^+. This formation of the phosphate anhydride by an enzyme of the cytosol is termed *substrate-level phosphorylation*, distinguishing it from oxidative phosphorylation. Conversion of the phosphate anhydride to a carboxylic acid is then

(cont. on p. 175).

These form the carboxyl groups of the pyruvate

Red and black symbols show how the carbons distribute into the 3-carbon compounds leading to pyrivate formation

Glucose

Hexokinase

Glucose 6-phosphate

Frutose 6-phosphate

Phosphofructokinase

Fructose 1,6-bisphosphate

Aldolase

Glyceraldehde phosphate

Dihydroxyacetone phosphate

NAD^+

NADH

Glyceraldehde-phosphate dehydrogenase

1,3-Bisphospho-glycerate

3-Phospho-glycerate

Enolase

2-Phospho-glycerate

By a similar series of reactions

is formed from di-hydroxy-acetone phosphate

Pyruvate kinase

Phosphoenol-pyruvate

Pyruvate

A

(cont. from previous page)

combined with the conversion of ADP to ATP. After the move of the phosphate to carbon 2, enolase, by removing water, yields phosphoenolpyruvate, which is then involved in a reaction in which a further molecule of ATP is formed from ADP. For each mole of glucose oxidized, the yield of ATP from glycolysis is 2 mol of ATP (4 mol of ATP produced, 2 mol of ATP utilized). In addition, 2 mol of NADH are produced (1 mol for each of the three-carbon products), each of which may yield a further 2 mol of ATP if oxidized by the electron transport chain (see p. 188B) via the glycerol phosphate shuttle.

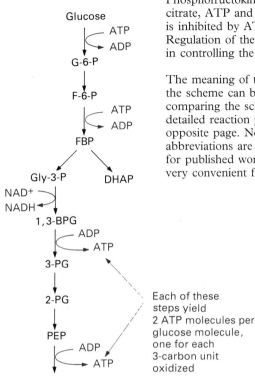

Phosphofructokinase is inhibited by citrate, ATP and NADH. Pyruvate kinase is inhibited by ATP and NADH. Regulation of these enzymes is important in controlling the rate of glycolysis.

The meaning of the abbreviations used in the scheme can be determined by comparing the scheme with the more detailed reaction pathway shown on the opposite page. Note that these abbreviations are not those recommended for published work. They are, however, very convenient for everyday use.

Each of these steps yield 2 ATP molecules per glucose molecule, one for each 3-carbon unit oxidized

B

Further metabolism of pyruvate

The pyruvate produced by the glycolytic pathway (and by other metabolic pathways) can be converted to oxaloacetate by the enzyme *pyruvate carboxylase*. Alternatively, in an oxidative reaction utilizing coenzyme A and NAD$^+$, it can be converted to acetyl CoA.

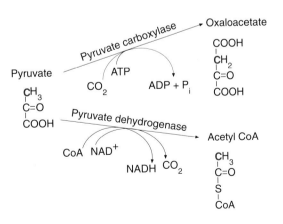

Pyruvate carboxylase has an absolute requirement for the presence of acetyl CoA as an activator. Biotin is also a coenzyme.

Pyruvate dehydrogenase exhibits product inhibition. Its activity is inhibited by a high acetyl CoA/CoA ratio.

Thiamin pyrophosphate is a coenzyme for pyruvate dehydrogenase.

A

Regulatory processes

Pyruvate dehydrogenase is regulated by the action of a protein kinase and a phosphatase.

The protein kinase is an example of the class of protein kinases that are not cAMP-dependent.

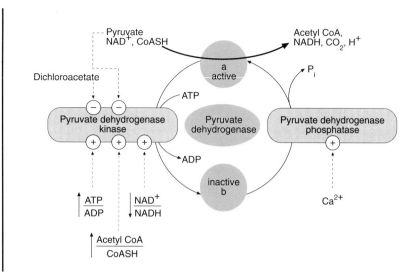

Pyruvate dehydrogenase is regulated by a kinase, that phosphorylates the enzyme, thereby inactivating it, and a phosphatase, that hydrolyses the phosphates and converts the enzyme to the non-phosphorylated, active form.

The kinase is activated by high ratios of ATP/ADP and acetyl CoA/CoA, and a decreased NAD^+/NADH ratio. Pyruvate inhibits the kinase. Activation of this kinase leads to inhibition of pyruvate dehydrogenase (see p. 181).

The compound dichloroacetate has been used as a drug to alleviate lactic acidosis. Its mechanism of action appears to include inhibition of the kinase which phosphorylates pyruvate dehydrogenase. As phosphorylation inhibits pyruvate dehydrogenase, the action of dichloroacetate is thus to accelerate conversion of pyruvate to acetyl CoA, thereby reducing formation of lactate.

B

The non-equilibrium reactions of glycolysis are catalysed by hexokinase, phosphofructokinase and pyruvate kinase. These are regulated by a number of allosteric modifiers.

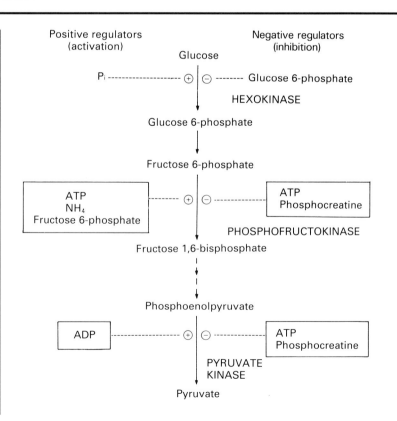

A

Phosphofructokinase (PFK-1) is also subject to positive regulation by *fructose 2,6-bisphosphate* (F-2,6-P_2), the concentration of which is controlled by a kinase and phosphatase, which are different forms of the same 49 kDa protein. These forms are the non-phosphorylated and phosphorylated forms of the protein, interconverted by the cAMP-activated protein kinase and a protein phosphatase.

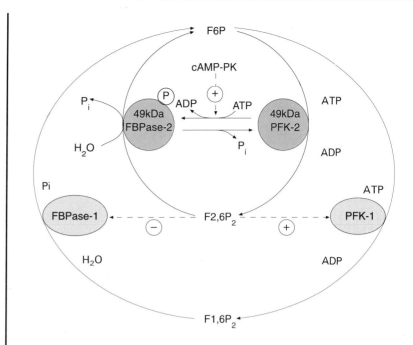

Fructose 1,6-bisphosphate (F1,6P_2) is hydrolysed to fructose 6-phosphate (F-6-P) by fructose 1,6-bisphosphatase (FBPase-1). This enzyme is inhibited by high concentrations of F-2,6-P_2, and thus activated by glucagon which lowers the F-2,6-P_2 concentration. F-2,6-P_2 potentiates the action of AMP on phosphofructokinase and fructose 1,6-bisphosphatase, on which AMP has similar actions to F-2,6-P_2.

B

Activities of the enzymes of glycolysis can be assayed in the homogenates of the disrupted cells of various tissues.

Phosphofructokinase is rate-limiting in all tissues.

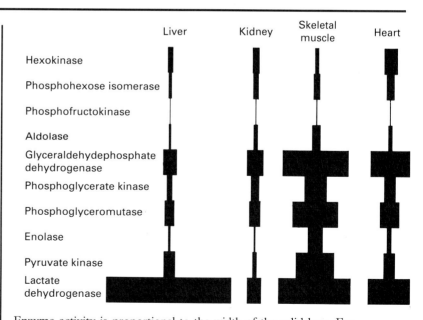

Enzyme activity is proportional to the width of the solid bars. For reference, the activity of kidney aldolase was 2 μmol min^{-1} per g fresh weight of tissue.

A

The citric acid cycle

Oxaloacetate and acetyl CoA can condense to form citrate.

The citrate that is formed is a pro-chiral molecule (see p. 10). For the structure of citrate see the next page.

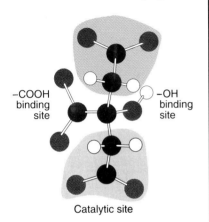

−COOH binding site

−OH binding site

Catalytic site

If −COOH and −OH binding sites are held on a surface, the two −CH$_2$COOH groups are distinguishable.

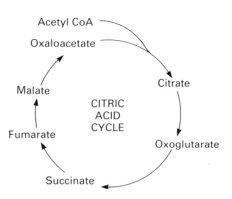

Citrate can be converted to oxaloacetate in a series of reactions in which CO$_2$ is lost and hydrogen removed. The oxaloacetate formed from the citrate can then react with a further molecule of acetyl CoA, and the sequence repeated an infinite number of times. In other words, the oxaloacetate is behaving as a catalyst, being regenerated in unchanged form after the reaction. This sequence of reactions, shown in detail on the next page, occurs in virtually every type of cell, with rare exceptions such as the mature erythrocyte.

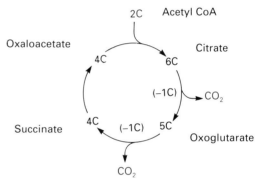

The net result of the reaction is the conversion of the acetyl residue to two molecules of CO$_2$ with delivery of eight hydrogen atoms to the acceptor molecules FAD and NAD$^+$.

B

The intermediates are free to enter and leave the cycle at any point, in which case there will be stoichiometric conversion of molecules, one into the other, in the normal way.

As an example, glutamate can be converted into aspartate by transamination with oxaloacetate to form aspartate and oxoglutarate, which then enters the citric acid cycle to be converted to oxaloacetate, which undergoes transamination with glutamate to yield further molecules of aspartate and oxoglutarate until an equilibrium position is reached.

1. Glutamate + Oxaloacetate ⟶ Aspartate + Oxoglutarate
2. Oxoglutarate $\xrightarrow[\text{Citric acid cycle enzymes}]{}$ Oxaloacetate

Step 1 then repeats with a further molecule of glutamate.
The net result is the stoichiometric conversion of glutamate to aspartate.

The citric acid cycle
—cont.

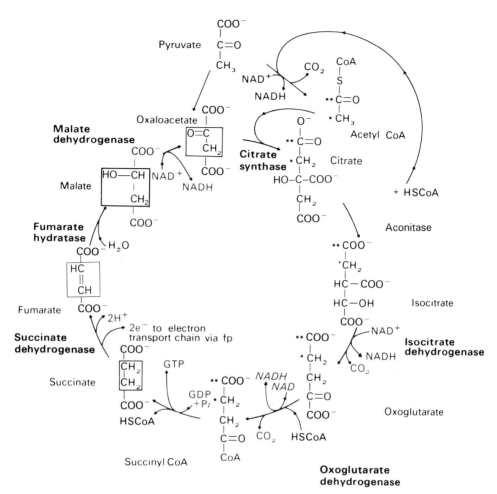

Note the reaction sequence

A similar series of reactions is also found in fatty acid oxidation.

The asterisks denote the carbons from acetyl CoA. The pro-chiral nature of citrate enables the stereospecific oxidation of the molecule. CO_2 from acetyl CoA is not released by oxoglutarate dehydrogenase during the first turn of the cycle. At succinate, the molecule becomes completely symmetrical, and $-COOH$ groups can then no longer be distinguished. The acetyl CoA label thus becomes randomized by the time oxaloacetate is formed, and during the next turn of the cycle 50% of the carbon from acetyl CoA will be released.

A

Regulation of the citric acid cycle.

Citrate synthase, isocitrate dehydrogenase and oxoglutarate dehydrogenase catalyse non-equilibrium reactions.

Both isocitrate dehydrogenase and oxoglutarate dehydrogenase are activated by Ca^{2+}, which may be of importance in muscle in relation to contraction stimulated by Ca^{2+}.

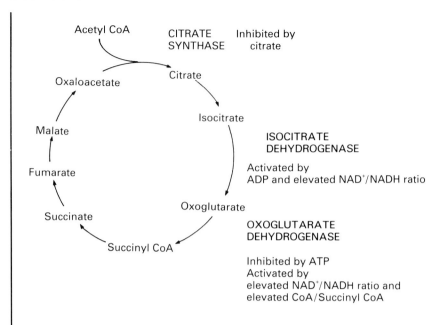

B

Pathway interactions

Major oxidative pathways may also act as routes linking other pathways.

The citric acid cycle links carbohydrate and fat metabolism, and is involved in the metabolism of amino acids. As is described in later sections, it provides an important control point for regulating metabolite flow in these pathways.

Succinyl CoA is used in porphyrin synthesis (see p. 166B).

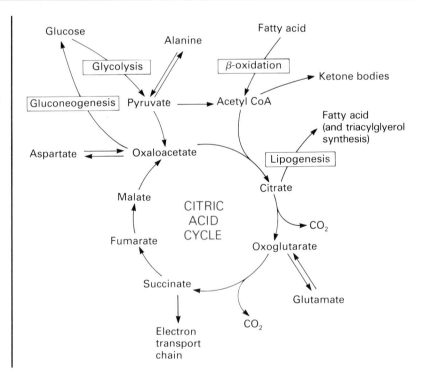

A

The enzymes of the glycolytic pathway link up the metabolism of glucose and other carbohydrates with fat and amino acid metabolism.

Substrates in boxes are important in interactions of the glycolytic pathway with other pathways.

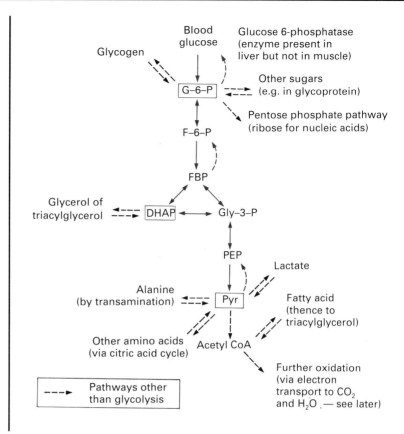

B

The glucose/fatty acid cycle

A reciprocal relationship exists between the rates of oxidation of glucose and fatty acids by muscle. This has important consequences when liver glycogen stores are depleted, and fatty acid oxidation can spare the use of glucose.

The mechanism by which fatty acids inhibit utilization of glucose involves a number of enzymes, in particular pyruvate dehydrogenase and phosphofructokinase.

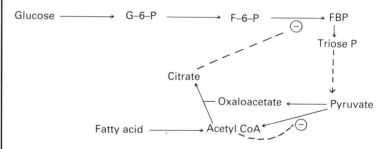

The most direct action of fatty acid oxidation is that it increases the level of acetyl CoA, and in the process utilizes CoA. An increase in the acetyl CoA/CoA ratio strongly inhibits pyruvate dehydrogenase. More indirectly, levels of citrate are elevated, leading to inhibition of phosphofructokinase.

A

The electron transport chain—*oxidation of NADH and succinate*

NADH, produced by the action of citric acid cycle enzymes, diffuses to the electron transport chain, where the first step in its oxidation is the removal (by NADH dehydrogenase) of a hydrogen atom and an electron. Succinate dehydrogenase accepts two hydrogen atoms (equivalent of $2H^+$ and $2e^-$) from succinate.

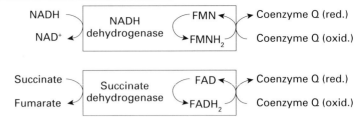

NADH dehydrogenase and succinate dehydrogenase are flavoproteins, with a molecule of FMN (NADH dehydrogenase) or FAD (succinate dehydrogenase) as a tightly bound coenzyme. This is reduced by transfer of reducing equivalents from NADH or succinate, and in its turn transfers these to coenzyme Q.

Coenzyme Q, CoQ
(Ubiquinone) oxidized form

Coenzyme Q (*ubiquinone*) is also the hydrogen acceptor for other flavoproteins such as are involved in β-oxidation and the oxidation of glycerol 3-phosphate. It is a lipid-soluble compound which can diffuse through the hydrophobic core of the membrane lipid bilayer.

(Ubisemiquinone)

Reduced coenzyme Q
(dihydroubiquinone ubiquinol)

It also has the property that it can form a semi-quinone, which may facilitate the transition between the two-electron transfer reactions of NADH and succinate dehydrogenases, and the cytochromes of the electron transfer chain with which ubiquinone reacts (see p. 186).

B

Ubiquinol is oxidized by a chain of *cytochromes*. The cytochromes are enzymes that have, as a prosthetic group, a haem molecule within which is bound an iron ion. Originally they were designated cytochromes *a*, *b* and *c*, and were detected using a spectroscope, exploiting the characteristic absorption of their haem groups. Other cytochromes have since been discovered, including c_1 and a_3.

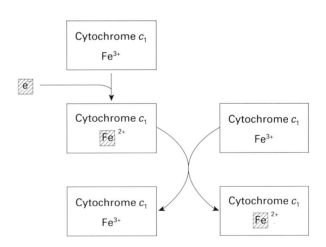

In the haem groups the iron is bound to a porphyrin similar to that found in haemoglobin. In the cytochrome haem, however, the iron undergoes oxidation and reduction between the ferric and ferrous forms, as electrons pass from one cytochrome to another. The flow of electrons is driven by the free energy change over the span of the electron transport chain.

A

The cytochromes are components of complex assemblies of protein subunits harbouring one or more haem molecules, together with iron–sulphur complexes (so-called *non-haem iron*) or, in the case of cytochrome *c* oxidase, copper ions.

The way in which the components of the chain interact can be investigated by isolating and partially solubilizing the inner mitochondrial membrane and then separating out the components. When this is done, four intact complexes can be purified.

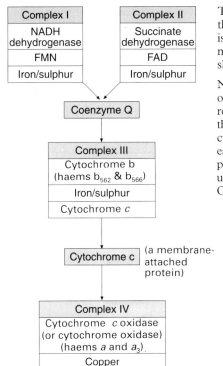

Complex I   NADH:ubiquinone oxidoreductase
Complex II   Succinate:ubiquinone oxidoreductase
Complex III   Ubiquinol:cytochrome *c* oxidoreductase
Complex IV   Ferrocytochrome *c* oxygen oxidoreductase

The available evidence indicates that the four components isolated from the mitochondrial membrane react in the sequence shown.

NADH and succinate are oxidized by complexes I and II, respectively. Electrons then pass through complex III, cytochrome *c* and complex IV, each component oxidizing the preceding component in turn, until complex IV is oxidized by O_2.

The arrows indicate the flow of electrons.
Note: The succinate dehydrogenase activity of complex II is the enzyme referred to as a component of the citric acid cycle (page 179).

B

Structural aspects of electron transport

The mitochondrion has an inner compartment, known as the *matrix,* surrounded by a membrane termed the *inner membrane*. Most of the enzymes of the electron transport chain are integral proteins of the inner membrane.

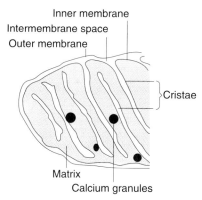
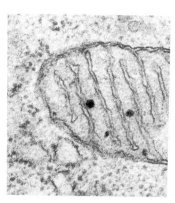

Surrounding the entire mitochondrion there is an *outer membrane* and the space between the inner and outer membranes is termed the intermembrane space. The inner membrane invaginates into the matrix to give areas of intermembrane space permeating the central areas of the mitochondrion. The inner areas of intermembrane space are known as *cristae*.

The inner membrane is not freely permeable, but contains transport enzymes for translocating ATP for ADP, and transporting anions. NADH or other nicotinamide nucleotides cannot cross the inner membrane.

The outer membrane is freely permeable to ATP, ADP and other small molecules.

Free energy of electron transport

As with any enzyme reaction, the reaction of a cytochrome with another component of the electron transport chain involves a change in free energy. Indeed, the release of free energy as a result of the oxidation of the hydrogen atoms of substances such as NADH or succinate, mediated through the cytochromes, yields the potential for the phosphorylation of ADP to ATP.

The reactions taking place in the electron transport chain are oxidation–reduction (redox) reactions. Redox reactions can be demonstrated in vitro by means of electrodes connecting solutions of the reactants, and the potential difference can be measured between the solutions in the two half-cells.

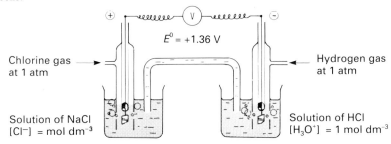

$E^0 = +1.36$ V

Chlorine gas at 1 atm

Hydrogen gas at 1 atm

Solution of NaCl
$[Cl^-]$ = mol dm^{-3}

Solution of HCl
$[H_3O^+]$ = 1 mol dm^{-3}

If such reactions are carried out at standard temperature (298 K) and pressure (1 atm), using 1 M solutions of reactants, the electronmotive force, as represented by the voltage recorded on the voltmeter, is referred to as standard oxidation–reduction potential of the reaction, or standard redox potential (E'_0). By convention, the standard redox potential of the H^+:H_2 couple (1 M H^+ in equilibrium with H_2 gas at 1 atm) is defined as 0 V, and all other redox potentials can be related to this. For the reactions of the electron transport chain the following table can be prepared:

Oxidant	Reductant	n	$E'_0(V)$
NAD$^+$	NADH + H$^+$	2	−0.32
Fumarate	Succinate	2	+0.32
Ubiquinone (oxidized)	Ubiquinone (reduced)	2	+0.10
Cytochrome b (3+)	Cytochrome b (2+)	1	+0.07
Cytochrome c (3+)	Cytochrome c (2+)	1	+0.22
Cytochrome a (3+)	Cytochrome a (2+)	1	+0.29
O_2 + 2H$^+$	H$_2$	2	+0.82

The standard free energy change $\Delta G^{\circ\prime}$ is related to $\Delta E'_0$, the change in the standard oxidation–reduction potential, by the equation

$$\Delta G^{\circ\prime} = nF \, \Delta E'_0$$

Where n is the number of electrons involved and F is the Faraday constant (96 500 coulombs). The units of $F \, \Delta E'_0$ are coulomb-volts, or joules (4.18 J = 1 cal). The overall span of the respiratory chain is 1.14 V. Thus the standard free energy change associated with oxidation of 1 mol of NADH can be calculated:

$$\Delta G^{\circ\prime} = \frac{-2 \times 96\,500 \times 1.14}{4.18} = -52.6 \text{ kcal mol}^{-1} = -220 \text{ kJ mol}^{-1}$$

A

Components of the electron transport chain

Cytochromes act as proton pumps.

Several of the components of the electron transport chain can link electron flow with the pumping of protons across the inner mitochondrial membrane from the matrix to the intermembrane space, thereby creating a proton gradient across the inner membrane.

NADH dehydrogenase, the cytochrome bc_1 complex and cytochrome c oxidase pump protons from the matrix to the intermembrane space. The face of the membrane to which protons are pumped is sometimes referred to as the *P-side* (P signifying positive), the face from which they are pumped being the N (or negative) *side*.

Inhibitors of electron transport are marked in red and inhibit complexes as indicated by the red bars

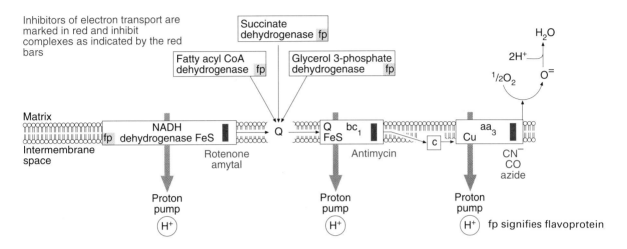

B

A schematic drawing of complex III, which bears cytochromes b and c_1 with non-haem iron.

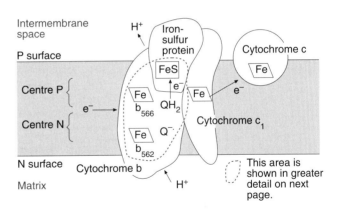

The complex actually contains two cytochromes b, designated b_{562} and b_{566} because of their spectral characteristics. The complex also contains a bound molecule of ubiquinone, which is thought to be involved in the transmission of electrons and removal of protons. The so-called Q-cycle has been proposed for this mechanism and is discussed in the next section.

The protonmotive Q-cycle

The Q-cycle is shown in detail as it does not lend itself to simple explanation. Essentially, however, it is a mechanism whereby electrons received in pairs by ubiquinone from NADH or succinate can be passed on singly to cytochrome c_1, *via* FeS. At the same time the protons are released into the intermembrane space.

Only enthusiasts need explore this figure!

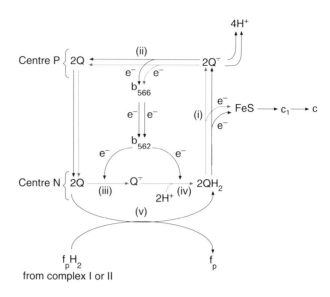

It has been recognized for many years from the results of kinetic experiments in which the rates of cytochrome oxidation and reduction were measured that cytochrome b does not directly transmit electrons to cytochrome c_1. As a result of these and other experiments, a *protonmotive* Q-cycle has been proposed as being consistent with the observed events. To envisage this cycle, one must follow the path of two molecules of ubiquinol (QH_2) entering the cycle simultaneously. One of these would have been produced by the action of complexes I and II, the other would have been regenerated by a full turn of the previous cycle. The two molecules of QH_2 first deliver one electron to cytochrome c_1 (*via* FeS), yielding the ubisemiquinone anion ($Q^{\bullet}$ as the product—stage (i). (Loss of an electron from QH_2 involves simultaneous loss of one proton to yield ubisemiquinone, $QH\bullet$, but this ionizes at physiological pH to $Q^{\bullet}$ and H^+. Thus, two protons are lost from QH_2 as a result of the transfer of one electron.) The protons lost as a result of this reaction are envisaged as being released in the intermembrane space as part of the proton pump mechanism of the complex. $Q^{\bullet}$ is then oxidized to ubiquinone by cytochrome b_{566}—stage (ii). One molecule of ubiquinone (Q) formed in this way is then reduced to ubisemiquinone (which once again exists in the anion form ($Q^{\bullet}$)) by one electron from b_{562} (which itself has been reduced by b_{566}—stage (iii). A second electron from b_{562} then further reduces $Q^{\bullet}$ to yield QH_2 (after accepting two protons, QH_2 being unionized at physiological pH)—stage (iv). The other molecule of Q formed at stage (iii) is reduced to QH_2 by complex I or II or other flavoprotein ubiquinone reductase. Thus, it may be envisaged that one molecule of QH_2 (following the red arrows) is regenerated (rather like a catalyst) and can enter the cycle again to facilitate the removal of two more hydrogens coming in via QH_2 produced by complex I or II, and so the cycle continues.

A

Structural aspects of mitochondria

Complex IV – cytochrome c oxidase.

Cytochromes are large assemblies composed of a number of subunits. This is illustrated by the detailed structure of cytochrome *c* oxidase. It is comprised of 13 subunits. Subunits I, II and III, which form the catalytic core of the complex, are encoded and synthesized in the mitochondrion, while the remaining 10 are nucleus encoded.

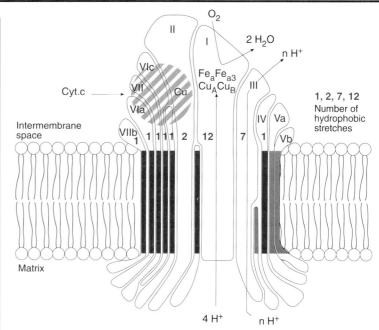

Roman numerals indicate different subunits. Subunits I and II between them carry the haem groups that interact with molecular oxygen, and the copper ions. The suggested site for binding cytochrome *c* is shown. The nucleus-encoded subunits may be involved in regulating the activity of the enzyme in response to changes in the metabolic state of the cell. The number of hydrophobic stretches (in arabic numerals) indicates how many times the protein traverses the membrane (compare with receptors — p. 259A and B).

B

The proton gradient that builds up across the inner membrane as a result of the pumping of electrons by complexes I, III and IV creates a force that drives the synthesis of ATP by an enzyme (ATP synthase) located on the inner surface of the mitochondrial inner membrane.

The *elementary particles* that contain the ATPase are readily visualized by the electron microscope, using the technique of negative staining at high magnification.

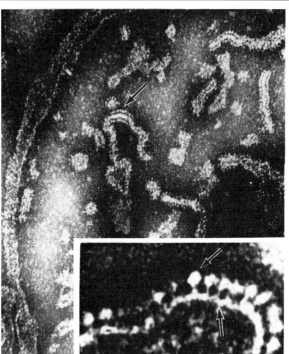

Electron micrograph of beef heart mitochondrion embedded in a thin phosphotungstate layer, showing elementary particles (arrowed).

A

ATP synthase.

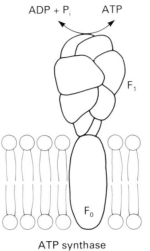
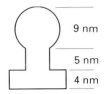

9 nm

5 nm

4 nm

Estimated size of F_0/F_1

ATP synthase

ATP synthase has been purified and found to be composed of a number of subunits. One group of these (termed F_1) comprises the catalytic unit (in the 'knob') and a section, or 'stalk', joins this unit to a complex (termed F_0) embedded in the membrane–lipid bilayer. Under certain conditions F_1 can function as an ATPase, hydrolysing ATP, and was often formerly referred to as the mitochondrial ATPase. Three heterodimers of $\alpha\beta$ subunits comprise the knob and the stalk is composed of the γ, δ and ϵ subunits.

B

Oxidative phosphorylation

The process whereby the oxidation of substrates by oxygen, through the cytochrome system, brings about the phosphorylation of ADP to ATP is known as *oxidative phosphorylation*.

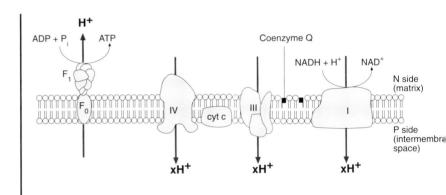

The pumping of protons by complexes I, III and IV to the outside (P-side) of the inner membrane creates a *protonmotive force* as a result of the proton gradient and the associated electrical potential that is built up across the membrane by the increase in positive charge. This *electrochemical potential* is maintained by the constant pumping of protons across the membrane, and tends to be dissipated by the action of ATP synthase, because this enzyme utilizes the electrochemical energy to drive ATP synthesis, with concomitant flow of protons back to the matrix. Each of the pumping sites (complexes I, III and IV) pumps sufficient protons to support the synthesis of approximately one molecule of ATP for each pair of electrons passing along the chain. Each pair of electrons reduces one oxygen atom ($\frac{1}{2}O_2$). This stoichiometry can therefore be expressed as a *P/O ratio* (number of molecules of inorganic P per $\frac{1}{2}O_2$).

When NADH is oxidized, with all three pumping sites functional, the P/O ratio is accordingly approximately 3. It is possible to block the chain with antimycin, and demonstrate that complex I can support oxidative phosphorylation with a P/O ratio of about 1. Succinate oxidation, when electrons pass only through two pumping sites (complexes III and IV) yields a P/O ratio of about 2. The exact stoichiometry of proton pumping and oxidative phosphorylation remains to be determined, but the above ratios are observed within limits of about 10%. Link this figure with 185A to observe the combined action of electron flow and proton pumping.

A

Respiratory control and uncoupling action

Some compounds, such as *dinitrophenol*, cause ATP synthesis to cease, and at the same time increase oxygen utilization. This is termed the *uncoupling* of ATP synthesis from electron flow, and such mitochondria are said to be *uncoupled*.

In coupled mitochondria, electron flow (oxygen utilization) is highly dependent on the relative amounts of ADP and ATP present, being high when ADP/ATP is high, and low when ADP/ATP is low. This is termed *respiratory control*.

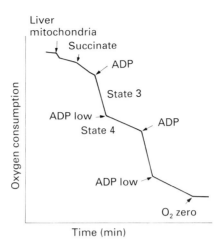

In liver mitochondria oxidizing succinate, addition of ADP elevates the ADP/ATP ratio (state 3) and increases oxygen utilization. When this ADP has been converted to ATP (low ADP/ATP ratio, state 4) oxygen utilization is low.

B

In the presence of uncoupling agents, electron flow is faster and there is little or no synthesis of ATP.

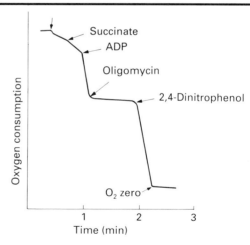

The trace from an oxygen electrode during oxidation of succinate by liver mitochondria shows the increased rate on adding ADP and arrest of oxidation by oligomycin (total inhibition of F_1 protein in *tightly coupled* mitochondria stops electron flow). Then addition of 2,4-dinitrophenol uncouples and allows rapid free flow of electrons. Tight coupling implies that electrons can only flow as ADP is phosphorylated to ATP, so that if this is totally inhibited by *oligomycin*, electron flow ceases. It is possible that 2,4-dinitrophenol uncouples by making the membrane leaky to protons by acting as a membrane-permeable proton carrier.

Flow chart of electron transport

The oxidation of 1 mol of glucose by the glycolytic pathway and the citric acid cycle generates 38 mol of ATP, or the ATP equivalent GTP.

The electron transport chain has been simplified in this diagram.

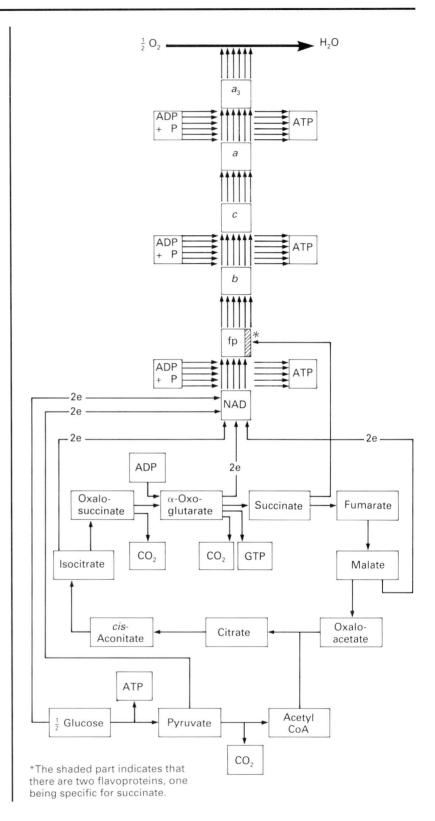

*The shaded part indicates that there are two flavoproteins, one being specific for succinate.

A

Translocases across mitochondrial membrane

A number of translocases exist in the inner mitochondrial membrane which transport a molecule into or out of the mitochondrion whilst simultaneously carrying another molecule in the opposite direction (*antiport;* see p. 277B).

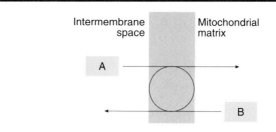

Pairs of compounds undergoing antiport translocation

A	B
Pyruvate	OH⁻
Phosphate	Malate
Citrate	Malate
Phosphate	OH⁻
ADP	ATP
Aspartate	Glutamate
Malate	2-Oxoglutarate

B

Transport of hydrogen from the cytosol into the mitochondrion can be mediated by glycerol 3-phosphate interacting with a dehydrogenase in the inner mitochondrial membrane.

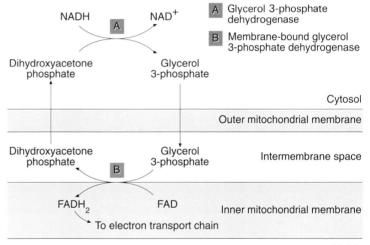

Hydrogen is first transferred from NADH in the cytosol to dihydroxyacetone phosphate, forming glycerol 3-phosphate by the soluble cytosolic glycerol 3-phosphate dehydrogenase. When the glycerol 3-phosphate is oxidized back to dihydroxyacetone by the mitochondrial membrane-bound enzyme, electrons enter the electron transport chain at cytochrome *b* via coenzyme Q, yielding only two molecules of ATP for each electron.

The overall reaction is thus the passage of hydrogens from the cytosolic NADH to the intramitochondrial electron transport chain.

Mitochondrial regulation of cell function

Over short time intervals, the components of each of the coenzyme systems

1. ATP + ADP + AMP
2. NADH + NAD$^+$
3. NADPH + NADP$^+$

sum to a constant total.

If one component of any of these three systems increases, it does so at the expense of the other component(s) of the system, i.e. if NADH increases, NAD$^+$ decreases, or if ATP increases, ADP or AMP (or both) will decrease. If this happens, the ratio NADH/NAD$^+$ or ATP/ADP will change.* Note. Within mitochondrion NADH/NAD$^+$ is in the range 0.1–0.2. In cytosol NADH/NAD$^+$ is about 0.001.

The rate of many different enzymes can be regulated by the change in the ratios ATP/ADP, NADH/NAD$^+$ and NADPH/NADP$^+$. Allosteric effects, and other effects on enzyme mechanisms, are involved in this regulation. The effects often originate in the mitochondrion, where respiratory control determines the relative concentrations of these coenzyme systems, and this can regulate citric acid cycle activity.

On the opposite page, there are some illustrations of the flow of metabolites in different situations in different types of tissue.

Situation (a): liver during fasting
During a fast, large amounts of fatty acid and oxaloacetate are entering the liver, the fatty acid for conversion to ketone bodies (see p. 202B) and the oxaloacetate to act as a precursor in gluconeogenesis (see p. 201A).

Situation (b): liver after a meal
Citrate produced by oxidation of excess nutrients is exported into the cytosol for fat synthesis.

Situation (c): brown adipose tissue
This tissue is important in thermoregulation. Rapid oxidation that does not lead to work or other forms of useful energy leads to the production of heat. If there is some means of removing respiratory control (such as by the constant hydrolysis of ATP) the effect will be similar to the uncoupling of oxidative phosphorylation; there will be unrestricted flow of electrons and oxidation of NADH, and a high rate of oxidative metabolism generating heat.

Situation (d): heart muscle
In cardiac muscle, constant contraction utilizes ATP, converting it to ADP. There are large numbers of mitochondria in heart muscle, ensuring that oxidation by the heart is essentially always aerobic.

*Note that the total amount of these nucleotides in the cell can change, over the course of hours, by changes in the rate that they are synthesized, or the rate at which they are broken down or utilized, for example for DNA synthesis. The interconversions referred to above occur on a much more rapid time-scale, over a period of minutes.

(a) Liver - during fasting

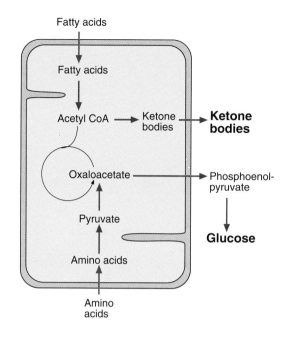

(b) Liver - after a meal

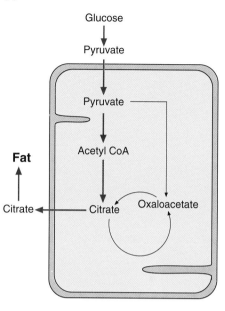

(c) Brown adipose tissue

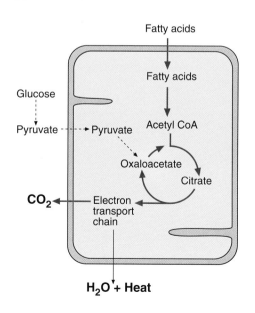

(d) Heart muscle

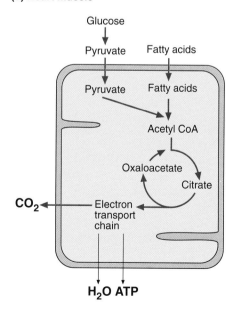

8

CARBOHYDRATE AND FAT METABOLISM

B. The fasting state— glycogenolysis, lipolysis, gluconeogenesis and ketogenesis

When an animal is deprived of food, it has one over-riding need—to maintain a supply of glucose to the blood. A fall in the blood glucose to below a critical level (about 2.5 mM in the human) leads to dysfunction of the central nervous system. In humans, this manifests as the symptoms of *hypoglycaemia* (muscular weakness and incoordination, mental confusion, sweating) leading, if the blood glucose falls further, to *hypoglycaemic coma*.

The immediate source of blood glucose is the liver glycogen store. *Glycogen phosphorylase* initiates this *glycogenolysis*, leading to secretion of glucose into the blood, a process that can be switched on with extreme rapidity (in seconds). In muscle, muscle glycogen is broken down to sustain muscular activity. On a longer time-scale (minutes), muscle protein is degraded to amino acids which are transported to the liver, where they are converted to glucose by the *gluconeogenic* pathways. The amount of glucose that must be produced by these pathways is lessened by the ability of tissues, apart from liver, to use *ketone bodies* as a fuel. These are formed in the liver from fatty acids (*ketogenesis*) which have been released from adipose tissue (*lipolysis*).

Release of hormones characteristic of the fasting state is crucial to this process. *Glucagon* activates glycogenolysis, gluconeogenesis and lipolysis. *Adrenaline* stimulates glycogenolysis and lipolysis, and release of glucagon.

A

Overview of the fasting state

During a fast, fat is mobilized and is the main substrate for oxidation in peripheral tissues, either as fatty acid or after conversion in liver to ketone bodies. Thus, the body needs to synthesize only minimal amounts of glucose from precious muscle amino acids, which in this state are the only long-term source of precursors for blood glucose synthesis.

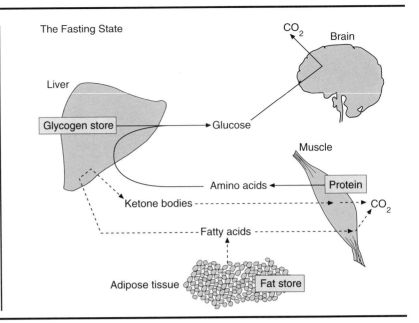

The Fasting State

B

In the fasting state:
1. Glucocorticoids release amino acids from muscle. In liver, gluconeogenic enzymes act to convert these amino acids to glucose.
2. Glucagon and adrenaline activate enzymes which release glucose from glycogen.
3. Fat mobilization is the result of *hormone-sensitive lipase* action on triacylglycerol in adipose tissue to cause release of fatty acids, which are converted to ketone bodies in liver. The lipase is activated by a cAMP-dependent protein kinase system (e.g. see p. 267).

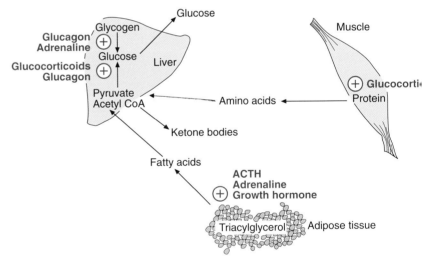

The Fasting State

Note: insulin is the only hormone that opposes the effect of these hormones of fasting state metabolism.

C

Many hormones act by stimulating adenylylcyclase.

Hormone	Tissues
Adrenaline	Muscle, liver, adipose tissue
Glucagon	Liver, adipose tissue
TSH	Thyroid
ACTH	Adrenals, adipose tissue

Appropriate receptors for the hormone must be present on the surface of the cell. In the absence of a specific receptor the hormone will not activate adenylylcyclase. TSH is thyroid stimulating hormone.

A

Details of metabolism in the fasting state

The pathways of major importance in the fasting state are *glycogenolysis*, *gluconeogenesis*, *ketogenesis* and ketone body utilization.

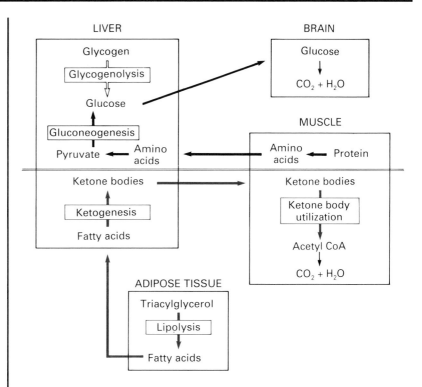

The system can be understood more easily by segregating it into the two main systems involved. Thus, the black arrows relate to the system for converting muscle protein to blood glucose, for use mainly by nervous tissue. The red arrows relate to the system for converting adipose tissue triacylglycerol to ketone bodies, used in place of glucose by tissues other than nervous tissue.

B

Glycogen and its degradation

Glycogen is a branched polymer of glucose. The main chain consists of α-(1→4) links between the glucose units. The branches are formed by α-(1→6) links, at approximately every 6 or 7 glucose units.

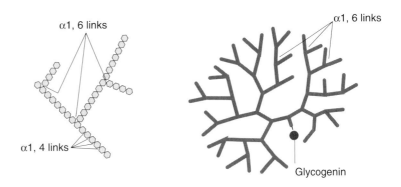

The glycogen chain is synthesized (see p. 214) by linking an additional glucose at its C-1 position to C-4 of an existing chain of glucose units. Thus glycogen gives a poor colour reaction with reagents that react only with free C-1 reducing groups, since all C-1 groups are engaged in bonding either with a C-4 group or with glycogenin (see page 215).

A

The formation of glucose from glycogen (glycogenolysis)

The diagram shows the detail of the α-(1→4) and α-(1→6) links in the glycogen molecule.

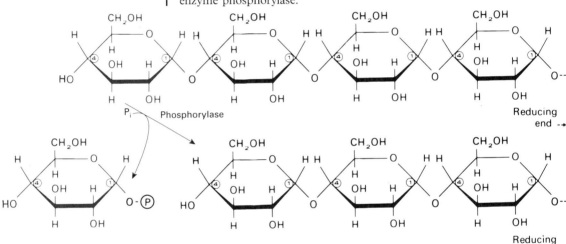

B

When glycogen is degraded, terminal α-(1→4) links are broken by the enzyme phosphorylase.

Phosphorylase

Reducing end --→

Reducing end --→

Note: 1. The product is glucose 1-phosphate
2. The phosphate is derived from inorganic phosphate, *not* ATP.
α-(1→6) links must be broken by other enzymes before phosphorylase can pass the branch point.

C

Degradation of glycogen to free glucose involves four enzymes:
1. Phosphorylase
2. 'Debranching enzyme'
3. Phosphoglucomutase
4. Glucose 6-phosphatase.

'Debranching enzyme' activity is carried out by maltotriosyltransferase (enzyme B in diagram) and 1,6-glucosidase (enzyme C).

Debranching enzyme

Phosphorylase (enzyme A) releases G-1-P. This must be converted to G-6-P, which is then hydrolysed to glucose.

Glucose 1-phosphate ──────────→ Glucose 6-phosphate
                       Phosphoglucomutase

                                  ──────────→ Glucose
                                  Glucose 6-phosphatase

Glucose 6-phosphatase is present in liver. It is not present in muscle. Thus, liver glycogen can act as a source of blood glucose, but muscle glycogen does not directly provide blood glucose (but see p. 199C and 202A).

A

Regulation of phosphorylase

After activation of a receptor linked to adenylylcyclase, cAMP activates the cAMP-activated protein kinase (PKA). This then activates, by phosphorylation, glycogen phosphorylase (GP) kinase, which in turn converts GPb to GPa.

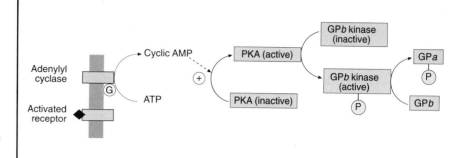

See p. 267B for details of the activation of PKA.

B

Changes of state of phosphorylase.

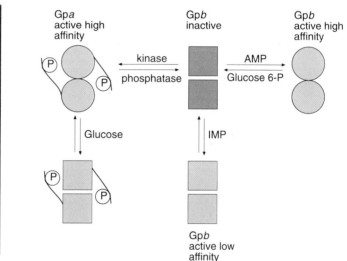

In the absence of effectors, GPb is a dimer composed of two identical subunits each of M_r 97 434. GPb is dependent on AMP for activity, and is inhibited by glucose 6-phosphate and ATP. Phosphorylation on Ser-14 by GPb kinase converts it to GPa, which adopts the R-form and this can be reversed by protein phosphatase 1 (see. p. 69). Glucose, which is an inhibitor, converts this to the T-form. Inosine 5'-phosphate (IMP) is a weak activator of GPb.

T-forms are shown as squares and R-forms as circles. Active states are shaded pink.

C

The de novo formation of glucose (gluconeogenesis)

The first step in the conversion of muscle amino acids to glucose is the removal of the amino group.

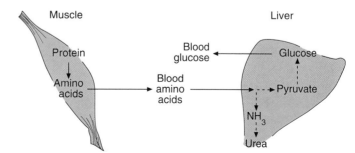

Phosphorylase can be activated within seconds by hormones such as adrenaline and glucagon, and possibly by nervous impulses, to meet the needs of the body for the immediate supply of glucose to the blood. However, the liver glycogen stores are sufficient to provide glucose for a few hours only. Thereafter, blood glucose can only be replenished by the conversion of muscle protein to glucose, after transport to the liver as amino acid.

A

The amino group is converted to urea by the *urea cycle*, which has increased activity during gluconeogenesis, giving an elevated urinary concentration of urea indicative of the protein breakdown that is occurring. See p. 155.

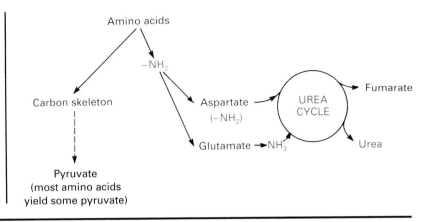

B

Substances other than amino acids may be used as substrates for gluconeogenesis.

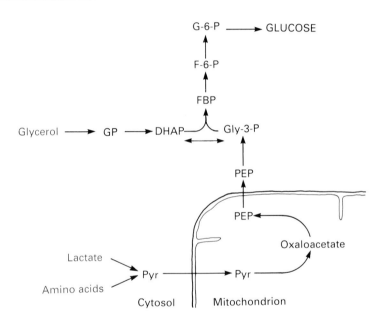

The major substrates for gluconeogenesis are shown in red. GP, glycerol 3-phosphate (see p. 191 and 213).

Only alanine directly provides pyruvate after removal of the amino group. The carbon skeletons of other amino acids must undergo a series of metabolic conversions to yield pyruvate. There is some evidence that this may be partly carried out in muscle, after which the pyruvate is converted to alanine by transamination, and delivered to the liver where it readily yields pyruvate after another transamination step.

Lactate provides pyruvate directly after oxidation by lactate dehydrogenase. Glycerol can be converted to glucose if it is first phosphorylated (by glycerol kinase) to glycerol phosphate, which can be oxidized to dihydroxyacetone phosphate by glycerol phosphate dehydrogenase.

A

Key enzymes in gluconeogenesis

The enzymes of particular importance in gluconeogenesis are located at the beginning and at the end of the pathway. Two ATP equivalents are required by the first two enzymes that convert pyruvate to phosphoenolpyruvate (PEP).

Control of the gluconeogenic pathway is primarily exercised at the steps involved in the formation of oxaloacetate and PEP. Elevated levels of acetyl CoA stimulate pyruvate carboxylase. The levels of pyruvate carboxylase and PEP carboxykinase are raised in situations requiring elevated glucose synthesis.

De novo synthesis of glucose proceeds via PYRUVATE.

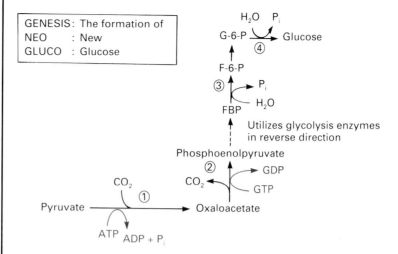

Four enzymes are of particular importance in gluconeogenesis:
1. *Pyruvate carboxylase.* This is the same enzyme as that which provides oxaloacetate for citric acid cycle activity. It is located within the mitochondrion.
2. *Phosphoenolpyruvate (PEP) carboxykinase* converts oxaloacetate to phosphoenolpyruvate. This enzyme is partially located in the mitochondrion, partially in the cytosol, and the location varies with the species of animal. It is most likely that oxaloacetate is converted to PEP in the cytosol in humans.
3. *Fructose bisphosphatase* is a hydrolytic enzyme in the cytosol.
4. *Glucose 6-phosphatase* is a hydrolytic enzyme in the membrane of the endoplasmic reticulum.

B

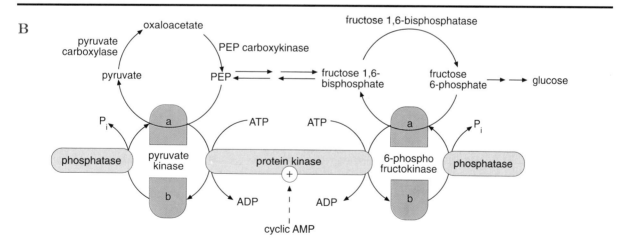

Pyruvate kinase and 6-phosphofructokinase are both subject to covalent control by protein kinases. In each case, action of the protein kinase converts the enzyme from an active form (a) to an inactive form (b), and these protein kinases are stimulated by cAMP, which thus inhibits glycolysis.

A

The Cori cycle

The lack of glucose 6-phosphatase in muscle prevents the formation of free glucose in that tissue. However, muscle metabolism can contribute to blood glucose indirectly. This has been termed the Cori cycle.

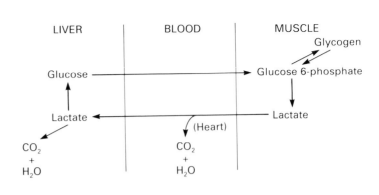

B

The formation of ketone bodies (ketogenesis)

Gluconeogenesis is usually accompanied by ketogenesis.

The ketone bodies comprise acetoacetate, β-hydroxybutyrate and acetone.

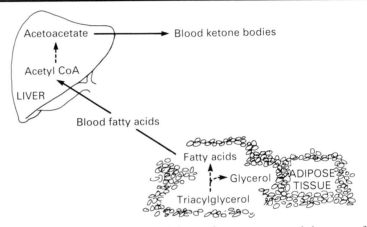

β-Hydroxybutyrate is the reduced form of acetoacetate and, because of the equilibrium position of the dehydrogenase that forms it, represents about two-thirds of the ketone bodies in the blood.

C

Three molecules of acetyl CoA are involved in the synthesis of acetoacetate. Acetoacetate (*not* acetoacetyl CoA) is the form secreted by liver into the blood.

The synthetic pathway for ketone bodies is found only in liver mitochondria but HMG CoA can also be synthesized in the cytosol for cholesterol synthesis.

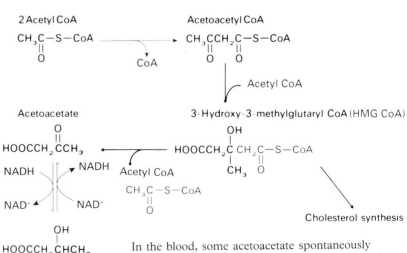

In the blood, some acetoacetate spontaneously decomposes to acetone, which is expired through the lungs and can be smelt in the breath of a diabetic suffering from severe ketosis.

A

The utilization of ketone bodies

Ketone bodies are produced in the liver, circulate in the blood and are utilized in muscle.

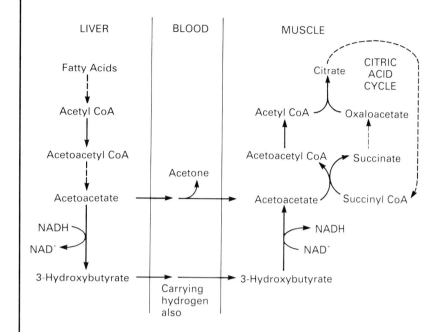

The availability of ketone bodies reduces to the minimum the amount of glucose which needs to be synthesized from muscle protein. The body need synthesize only that amount of glucose required by the brain and other cells of the central nervous system. Even these, during a prolonged fast, can utilize ketone bodies as a respiratory substrate, but it is still essential that the blood glucose level be maintained above about 2 or 3 mmol per litre.

B

Regulation of fatty acid oxidation and ketogenesis

Fatty acid oxidation takes place within the mitochondrion and the rate of oxidation is regulated by modulation of the rate at which fatty acids are transported across the mitochondrial membrane.

Fatty acids are transported across the mitochondrial membrane as esters of a quaternary ammonium hydroxyacid, *carnitine*. The carnitine acyl esters are formed in a reversible reaction catalysed by the enzyme *carnitine palmitoyltransferase*.

Carnitine   +   Palmitoyl CoA $\longrightarrow$ Palmitoyl carnitine   +   CoA

A

The role of carnitine palmitoyltransferase

Two forms of carnitine palmitoyltransferase exist in mitochondria, one in the outer and one in the inner membrane.

In liver mitochondria, carnitine palmitoyltransferase in the outer mitochondrial membrane is regulated by malonyl CoA, which inhibits it by interacting with a subunit. This, as shown in the figure, can exist in dissociated form. Incoming fatty acyl CoA thioesters are converted to acylcarnitine esters by this enzyme (CPT$_1$) followed by transport through the inner mitochondrial membrane by the acylcarnitine:carnitine antiporter (shown as an open cylinder). Carnitine palmitoyltransferase (CPT$_2$) in the inner membrane (which is not regulated by malonyl CoA) reforms acyl CoA thioesters in the internal space (matrix) and they then enter the β-oxidation pathway.

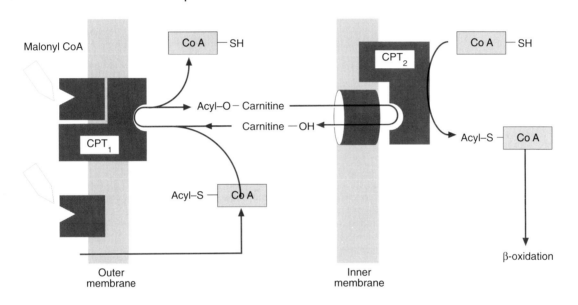

B

Control of the blood glucose level

During a fast, the liver can regulate with very great precision and rapidity the amount of glucose secreted into the blood, to match exactly that amount being removed. This ensures a constant blood sugar level during fasting.

The constancy of the fasting blood sugar level is to be contrasted with the way in which the blood sugar level rises on feeding.

The fasting blood sugar is maintained within precise limits. A fasting person can leap suddenly out of a chair, and run rapidly, with only minor effects on the blood glucose level, because of the speed with which the liver can increase or decrease glucose output.

In contrast, glucose entry from the gut, and glucose utilization by tissues, are not regulated with the same degree of precision. Thus, during ingestion of carbohydrates, the blood sugar rises.

A few hours in the history of John Smith's blood sugar

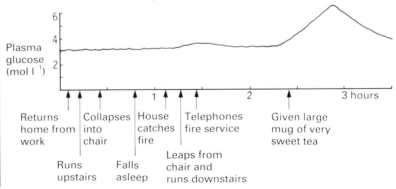

A

Disordered blood glucose levels

A number of pathological conditions occur in which regulation of the blood glucose level is abnormal. The most common of these is diabetes mellitus.

As explained in the previous section, in the fasting state the blood glucose normally maintains a remarkably constant concentration. In pathological states, it may be elevated *(hyperglycaemia)*, or depressed *(hypoglycaemia)*. Hyperglycaemia most commonly results from a deficiency in either the amount or the effectiveness of insulin, which is the basis of the condition known as diabetes mellitus. In this disease, elevation of the blood glucose is characteristically sufficient to exceed the renal threshold so that sugar appears in the urine *(glucosuria)*, and these phenomena form the basis of the initial investigation.

Two forms of diabetes mellitus are recognized:
1. Type 1—the so-called juvenile type, because it often manifests itself in children and young adults, and in which there is a marked failure to release insulin from the B-cells (also called β-cells) of the islets of Langerhans in the pancreas. The only effective treatment is administration of insulin, and this type is often referred to as insulin-dependent diabetes mellitus (IDDM). Type 1 diabetes is considered to arise in many patients as a result of an autoimmune attack on the B-cells by antibodies against this tissue formed by the patient's own immune system.
2. Type 2—the so-called maturity-onset type, because it often first manifests itself in the adult, especially if obese. However, as with type 1, distinctions are blurred as it can occur in younger individuals ('maturity-onset diabetes of the young', a variant characterized by an autosomal-dominant mode of inheritance). In type 2 diabetes there may be secretion of significant amounts of insulin. The reason for the failure to maintain a normal blood glucose level in cases where insulin is secreted is not well understood. In mild cases of type 2 diabetes, treatment initially takes the form of dietary recommendations and the use of drugs to stimulate release of insulin from the pancreas. Thus, it is often referred to as non-insulin-dependent diabetes mellitus (NIDDM). Ultimately, however, the administration of insulin by injection may be necessary, as it is in cases of type 1 diabetes.

B

The glucose tolerance test is used to assist in diagnosis of diseases that involve abnormal blood glucose levels. If insulin and free fatty acid levels are measured during the course of the test, they reflect the change in metabolic states from the fasting state, which is operating at the start of the test, to the fed state, to which the tissues have adjusted after about 1 hour.

The glucose tolerance test is carried out after an overnight fast. Glucose is then given orally, and blood samples are taken at intervals thereafter. The concentration of glucose and other substances is measured in these blood samples.

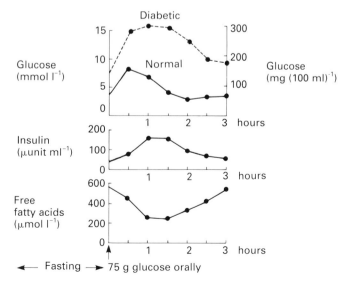

A

The classical acute diabetic crisis, in patients in whom blood insulin levels have reached critically low values, is characterized by:
1. A high blood glucose
2. Very high blood ketone bodies and free fatty acids
3. Loss of salts
4. Excessive urea excretion
5. Severe tissue wasting.

In the absence of insulin there is no regulator to keep the action of hormones such as the glucocorticoids, glucagon, ACTH and growth hormone within limits. Excessive production of glucose and ketone bodies occurs. The continual excretion of these by the kidney causes dehydration and loss of cations.

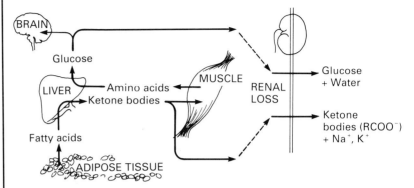

The severe tissue wasting and urea excretion of the diabetic crisis are due to the degradation of muscle protein as a result of the abnormally prolonged and extreme gluconeogenic response due to the action of gluconeogenic hormones, which are unrestrained in severe insulin deficiency, and de-amination of amino acids in liver with attendant urea synthesis. This leads to grossly elevated blood glucose levels in the acute diabetic state, regardless of whether food has been taken or not. Similarly, in this condition lipid is mobilized from adipose tissue, leading to greatly elevated blood fatty acid levels, and formation of large amounts of ketone bodies, elevated blood ketone levels, acetone in the breath and a massive excretion of ketone bodies in the urine.

B

Hyperinsulinaemia; excessive secretion of insulin.

Excessive secretion of insulin occurs in cases of *insulinoma*, a tumour of the pancreas. Severe hypoglycaemia may result on fasting, due to the excess insulin blocking the action of fasting state hormones.

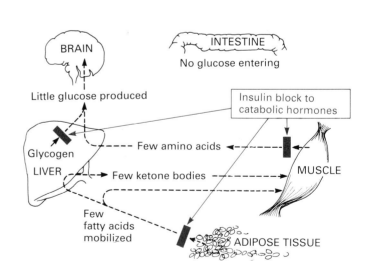

Note. The block of hormonal responses will probably not be total; the dashed lines indicate very low amounts of the relevant materials.

A

The role of lactate

Oxidation in the deficiency or absence of electron transport can continue as a result of the action of lactate dehydrogenase.

The red blood cell lacks mitochondria and derives its energy entirely from glycolysis. This requires the continual oxidation of NADH back to NAD^+, otherwise the enzyme glyceraldehyde-phosphate dehydrogenase would cease to function (see p. 174). This can be achieved by the enzyme lactate dehydrogenase, which catalyses the reaction

$$\text{Lactate} + NAD^+ \rightleftarrows \text{Pyruvate} + NADH + H^+$$

Thus the pyruvate formed by the glycolytic enzymes is converted to lactate which is secreted into the blood.

A similar sequence of reactions occurs in muscle when the provision of oxygen is insufficient to oxidize all of the NADH produced by glycolysis, and in other tissues when they are deprived of oxygen. This is the cause of elevated blood levels of lactate during severe exercise,[*] or as a result of tissue anoxia of pathological origin, e.g. after cardiac infarction.

The reactions involved can be summarized as follows:

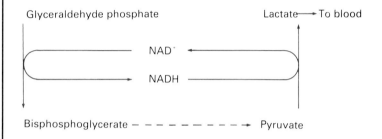

[*] Note that a sprinter is actually using about ten times as much oxygen during a sprint as he does whilst at rest. The lactate is produced as a result of the need to metabolize yet more substrate than even this amount of oxygen can provide for.

B

Specialization of metabolism in muscle

The fibres of a muscle can be classified into types according to their content of different enzymes, and this is related to their function.

The muscle fibres can be classed as type I or type II fibres according to their content of certain enzymes, such as the enzymes of the glycolytic pathway. Fibres containing high levels of these enzymes are classed as type II fibres, or fast-twitch fibres. These fibres contain relatively few mitochondria. Other fibres that contain lower levels of glycolytic enzymes but higher levels of citric acid cycle enzymes and cytochromes are classed as slow-twitch or type I fibres. Formerly, muscles containing large numbers of type I or type II fibres were referred to as red and white muscles, respectively (red muscles containing high amounts of myoglobin and cytochromes). Type II fibres can contract very rapidly, but for short periods of time only (lobster abdominal muscle, responsible for the tail flick escape reaction, is an example). Type I fibres on the other hand are capable or more sustained, slower contraction.

A

Glycogen storage disease

Hypoglycaemia resulting in acidosis.

The rare inborn error *type I glycogen storage disease* (von Gierke's disease) is an excellent model in which to study interrelationships between carbohydrate and fat metabolism, because the metabolic lesion is precisely known.

The specific defect is a deficiency of liver glucose 6-phosphatase. Over a period of years, large deposits of glycogen may form in the liver. On fasting for only a short period, there is severe hypoglycaemia, mild ketosis and mild *lactic acidosis*.

In this condition, the pancreas functions normally. Excessive ketosis does not result because a *basal* level of insulin is secreted that keeps fatty acid mobilization within reasonable limits.

Glucose 6-phosphatase deficiency

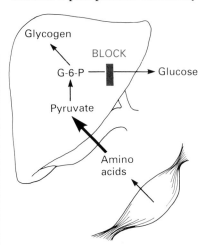

In the FASTING STATE:
1. Glucose formation by liver cannot occur, because:
   a. Glycogen breakdown to glucose is blocked
   b. Gluconeogenesis is blocked.
2. Hypoglycaemia results. Hormones are secreted in response to the hypoglycaemia, e.g. glucocorticoids, which cause excessive mobilization of muscle amino acids, which enter the liver.

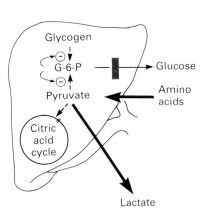

3. G-6-P levels build up and feedback inhibition blocks glycogen breakdown, and also the route from pyruvate to G-6-P.
4. Because of the hypoglycaemia and the fasting state of the body, citric acid cycle activity is reduced.
5. Thus, all normal routes for disposal of pyruvate are blocked.
6. This forces it towards lactate, which spills over into the blood (lactic acidosis).

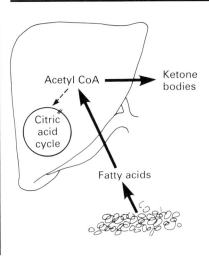

7. Hormones secreted in response to the hypoglycaemia cause release of fatty acid from adipose tissue.
8. Formation of ketone bodies results.

8

CARBOHYDRATE AND FAT METABOLISM

C. The absorptive state— synthesis of energy storage compounds

Ingested food will supply the immediate needs of an animal for precursors for essential synthesis of structural compounds and for oxidative substrates. It is not uncommon, however, in modern industrial societies, for food in excess of these needs to be eaten. For an animal in the wild, a meal may well come after a significantly long fast. In both of these cases, food that is surplus to immediate needs will be converted into the materials used as energy stores, that is, fat and glycogen. In the first case this will make a contribution to increasing obesity, while in the second case it will help to provide reserves to sustain the animal during any subsequent enforced fast. The processes operative during this metabolic state are thus *glycogenesis* and *lipogenesis*.

The high incidence of ischaemic heart disease, caused in part by the formation of plaques in the arteries *(atherogenesis)*, has directed much attention to the metabolism of cholesterol. In this, the *plasma lipoproteins* play a significant role. The only oxidative pathway for the excretion of cholesterol is its conversion to bile acids. Hence, increasing excretion of bile acids and inhibition of the pathway for cholesterol synthesis are therapeutic strategies for reducing *atherosclerosis*.

A

Overview of the fed state

After a meal, food which is surplus to immediate energy requirements is converted to glycogen and fat.

In this book we call this state the *fed state* of metabolism.

The fed state of metabolism operates whilst food is being absorbed from the intestine, and is more correctly termed the *absorptive state*.

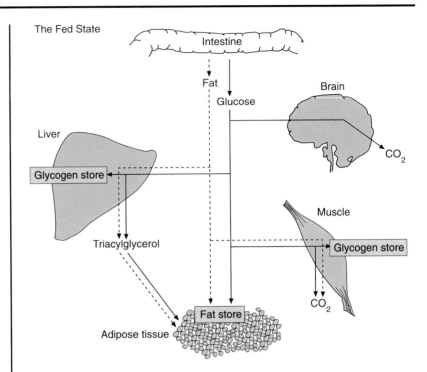

Carbohydrate and fat are oxidized to CO_2 and H_2O in peripheral tissues to drive synthetic reactions, and sustain cell function. Nutrients surplus to immediate requirements are laid down as fat or glycogen.

B

In the fed state, insulin is secreted, and promotes the synthesis of glycogen and triacylglycerol from glucose. It inhibits the release of fatty acids from adipose tissue triacylglycerol.

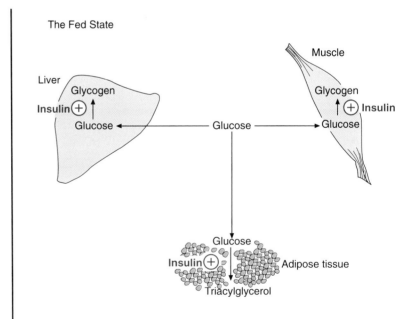

A

Digestive processes during the absorptive state

Before food can be absorbed from the digestive tract it must be processed to convert it to a form suitable for absorption.

Enzymes released during digestion hydrolyse food constituents.

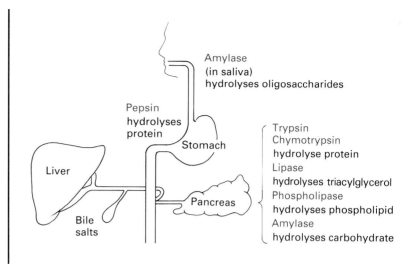

Amylase (in saliva) hydrolyses oligosaccharides

Pepsin hydrolyses protein

Stomach

Liver

Bile salts

Pancreas

Trypsin
Chymotrypsin
hydrolyse protein
Lipase
hydrolyses triacylglycerol
Phospholipase
hydrolyses phospholipid
Amylase
hydrolyses carbohydrate

B

Protein digestion.

Proteins are hydrolysed to smaller peptides by proteases such as pepsin, trypsin and chymotrypsin. These peptides may then be hydrolysed to amino acids by peptidase enzymes or absorbed as peptides.

Enzyme	Secreted by	Precursor	Converted to active enzyme by
Pepsin	Stomach	Pepsinogen	HCl
Trypsin	Pancreas	Trypsinogen	Enterokinase
Chymotrypsin	Pancreas	Chymotrypsinogen	Trypsin

Leucine amino-peptidase Pepsin Trypsin Chymotrypsin Carboxypeptidase A

$$CH_3\ CH_3$$

$$\underset{\text{Leu}}{H_2N-CH-\overset{\|}{\underset{O}{C}}NH}-\underset{\text{Phe}}{CH-\overset{\|}{\underset{O}{C}}NH}-\underset{\text{Lys}}{CH-\overset{\|}{\underset{O}{C}}NH}-\underset{\text{Ser}}{CH-\overset{\|}{\underset{O}{C}}NH}-\underset{\text{Tyr}}{CH-\overset{\|}{\underset{O}{C}}NH}-\underset{\text{Gly}}{CH_2-\overset{\|}{\underset{O}{C}}NH}-\underset{\text{Gly}}{CH_2-\overset{\|}{\underset{O}{C}}NH}-\underset{\text{Phe}}{CH-COOH}$$

	Alternative residues that permit enzyme action		
	Leu	Arg	Phe	Neutral amino acids
	Tyr			

The proteolytic enzymes act on the peptide bond on the carboxyl side of certain R-groups, as shown (carboxypeptidase acts on the amino side of appropriate C-terminal residues). The red arrows show which bond is hydrolysed.

A

Carbohydrate digestion.

Amylases in saliva and pancreatic juice break α-(1 → 4) bonds and hydrolyse amylose to free glucose, and maltose. Because α-amylases do not break α-(1 → 6) bonds, amylopectin is degraded to oligosaccharides known as dextrins.

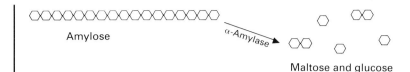

Amylose

α-Amylase

Maltose and glucose

Other enzymes are present in the gut, especially disaccharidases:

$$Sucrose \xrightarrow{Sucrase} Glucose + Fructose$$

$$Maltose \xrightarrow{Maltase} Glucose + Glucose$$

$$Lactose \xrightarrow{Lactase} Galactose + Glucose$$

Absence of disaccharidases occurs, either as an hereditary trait (when it may manifest itself in infancy) or as a result of infection. Such a deficiency may be the cause of chronic diarrhoea. This is especially so in the case of lactase. A deficiency of this enzyme gives rise to the condition known as lactose intolerance.

Cellulose cannot be degraded by mammalian gut enzymes. This and other plant polysaccharides (see p. 145B) which pass through to the lower bowel may have an important function in bowel action. They contribute to dietary fibre and are a component of 'roughage'.

B

Fat digestion.

Fats are hydrolysed to monoacylglycerol and fatty acid by pancreatic lipase, see page 211A. Other lipases include lipoprotein lipase, page 223, and hormone-sensitive lipase, page 196B.

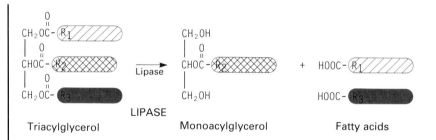

Triacylglycerol   Monoacylglycerol   Fatty acids

The coloured symbols represent the alkyl chains of the acyl residues.

Lipase acts on micelles of triacylglycerol and bile salts, i.e. the bile salts emulsify the fat droplets containing the triacylglycerol:

$$Large\ fat\ droplets \xrightarrow{Bile\ salts} Micelles\ of\ triacylglycerol\ and\ bile\ salts$$

C

Nomenclature of enzymes.

Enzymes that are involved in the degradation of macromolecules may be of two general types. (1) Enzymes that attack bonds in the interior of the macromolecule have the prefix 'endo-'. Thus, α-amylase is an *endo-amylase*, and trypsin is an *endo-peptidase*. (2) Enzymes that only break bonds at the terminals of the chains of the macromolecule bear the prefix 'exo-'. Thus, β-amylase, which is an enzyme that removes only the terminal maltose of amylose, is an *exo-amylase*, and carboxypeptidase is an *exo-peptidase*.

A

Detailed metabolism in the fed state

The fed (or absorptive) state involves the synthesis of glycogen and fat from blood glucose.

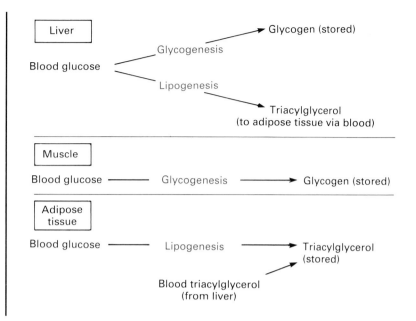

B

The pathways involved in the synthesis of glycogen and fat in liver can be divided into three groups.

Note that glycerol phosphate (glycerol 3-phosphate), derived from dihydroxyacetone phosphate, is an important link between fat and carbohydrate metabolism. It provides the glycerol backbone of many lipid molecules.

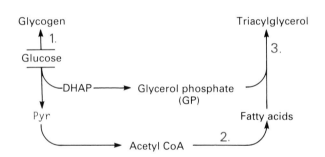

1. The formation of glycogen from glucose (glycogenesis)
2. The synthesis of fatty acid from acetyl CoA
3. The synthesis of triacylglycerol from glycerol phosphate and fatty acid. } (lipogenesis)

C

Glycerol 3-phosphate dehydrogenase interconverts dihydroxyacetone phosphate and glycerol 3-phosphate

$$
\begin{array}{c}
CH_2OH \\
| \\
CHOH \\
| \\
CH_2O\textcircled{P}
\end{array}
+ \; NAD^+
\;\rightleftharpoons\;
\begin{array}{c}
CH_2OH \\
| \\
C=O \\
| \\
CH_2O\textcircled{P}
\end{array}
+ \; NADH \; + \; H^+
$$

Glycerol 3-phosphate dehydrogenase

Glycerol 3-phosphate   Dihydroxyacetone phosphate

A

Biosynthesis of glycogen

Glycogen is synthesized by addition of glucose units to the (non-reducing) C-4 of the terminal glucose of a pre-existing α-(1 → 4) chain primer.

This is catalysed by glycogen synthase, and the glucose to be added reacts in the form of UDP-glucose.

Glycogen
synthase

UDP

B

Branches in the glycogen molecule are formed by breaking the growing α(1 → 4) chain and transferring the residues to form an α-(1 → 6) link.

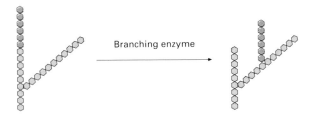

Branching enzyme

Glycogen branching enzyme transfers a chain of six glucose residues from the end of one growing chain of α-(1 → 4)-linked residues to form an α-(1 → 6) bond between C-1 of the glucose chain transferred and C-6 of a residue in another chain about 6 residues distant from a previous α-(1 → 6) link.

A

Initiation of glycogen synthesis.

Glucose residues are attached to a protein, glycogenin, to form a maltosaccharide primer, which is extended by glycogen synthase, using UDP-glucose and branching enzyme.

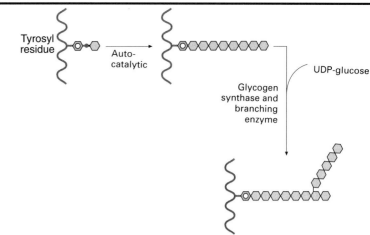

After the addition of a glucose unit to a tyrosyl residue of glycogenin by a mechanism yet to be established, the molecule becomes autocatalytic and a further seven glucose units are added to form the chain which primes the action of glycogen synthase. The current evidence is for one maltosaccharide chain to be grown on each glycogenin molecule. The linkage to the glycogenin tyrosyl is by the C-1 group.

The combination of all of these reactions results in the branched macromolecule, glycogen, shown on page 197B.

B

Glycogen storage diseases

An inherited deficiency of one or other of the enzymes involved in glycogen metabolism can lead to glycogen storage disease, some types of which are listed.

Type	Name	Enzyme affected	In which tissue	Remarks
I	Von Gierke's disease (see p. 208B)	Glucose 6-phosphatase	Liver Kidney	Low fasting blood glucose, enlarged liver with elevated glycogen levels
III	Limit dextrinosis	Debranching system★	Liver or muscle	Phosphorylase acts on α-(1 → 4) bonds until an α-(1 → 6) bond is reached. The resulting molecule is called a limit dextrin
V	McArdle's disease	Phosphorylase	Muscle	Muscular pain, weakness and stiffness after only mild exercise
VI		Phosphorylase	Liver	Has some features similar to type I but, as gluconeogenesis is not blocked, hypoglycaemia may not be so severe

★The deficiency can be either in the maltotriosyl transferase or 1,6-glucosidase activities (see p. 198C) and can be in liver or muscle or both.

A

Regulation of glycogen synthase

Glycogen synthase is regulated by its phosphorylation state in a complex way. It is phosphorylated on nine different serines by several kinases, and dephosphorylated by protein phosphatase 1.

Glycogen synthase (more active)

CAM kinase,
Glycogen synthase kinase,
cAMP-dependent kinase, and
phosphorylase kinase all
phosphorylate glycogen synthase

ATP

ADP

P_i

Protein phosphatase 1

Glycogen synthase (phosphorylated—less active)

Phosphorylation tends to inactivate glycogen synthase, whereas it activates phosphorylase, resulting in a concerted regulation of glycogen synthesis and breakdown. Insulin is thought indirectly to activate protein phosphatase 1.

B

Pentose phosphate pathway ('pentose shunt')*

This cytosolic pathway is found in many cell types.

The enzymes of the pentose phosphate pathway are important in providing NADPH for fatty acid synthesis. They also yield pentoses for nucleotide and nucleic acid synthesis.

Two dehydrogenases are involved in the production of this NADPH:
1. Glucose 6-phosphate dehydrogenase
2. 6-phosphogluconate dehydrogenase.

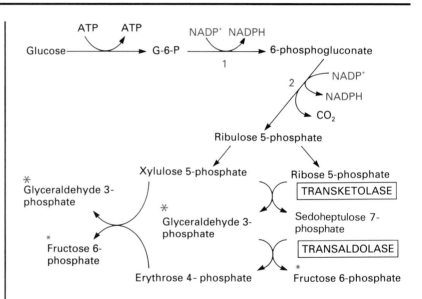

ATP   ATP $NADP^+$ NADPH

Glucose ⟶ G-6-P ⟶ 6-phosphogluconate
                        1

2        $NADP^+$
         NADPH
         CO_2

Ribulose 5-phosphate

Xylulose 5-phosphate   Ribose 5-phosphate

*Glyceraldehyde 3-phosphate   TRANSKETOLASE

*Glyceraldehyde 3-phosphate

Sedoheptulose 7-phosphate

*Fructose 6-phosphate

TRANSALDOLASE

Erythrose 4-phosphate   *Fructose 6-phosphate

* Enter other pathways such as the glycolytic pathway

A mnemonic for the pentose shunt is that, after formation of ribulose 5-phosphate, three pairs of reactants successively sum to ten carbons. Then a four-carbon compound and five-carbon compound (total of nine carbons) combine to give fructose 6-phosphate and glyceraldehyde 3-phosphate, which enter other pathways of metabolism.

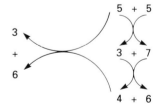

3
+
6

5 + 5

3 + 7

4 + 6

The term 'shunt' was coined rather misleadingly because at the time it was considered that one role of the pentose phosphate pathway was to shunt glucose away from glycolysis.

* Also known as the hexosemonophosphate pathway or hexosemonophosphate shunt.

Lipogenesis

The synthesis of triacylglycerol from glucose.

Lipogenesis is the synthesis of fat from non-fat materials, especially carbohydrate and includes synthesis of fatty acids and triacylglycerol

Steps in lipogenesis involve:
1. Conversion of glucose to acetyl CoA (via glycolytic pathway)
2. Formation of glycerol phosphate (via glycolytic pathway and reduction of dihydroxyacetone phosphate)
3. Synthesis of triacylglycerol from fatty acyl CoA and glycerol phosphate (see p. 220B for fatty-acyl CoA synthesis).

NADPH for lipogenesis is provided by the pentose shunt. Further NADPH is formed by the malic enzyme; cytosolic malate dehydrogenase and the malic enzyme provide a transhydrogenase mechanism in the cytosol, transferring hydrogen from NADH to NADPH.

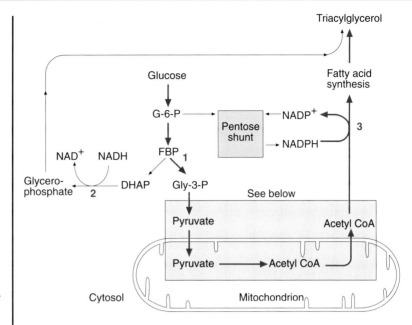

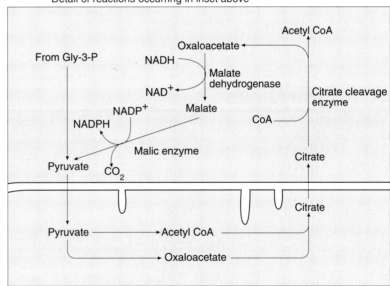

Detail of reactions occurring in inset above

Acetyl CoA is not transported readily through the mitochondrial membrane. It is converted to citrate, for which there is a membrane transporter to transport it to the cytosol. Citrate is in turn converted in the cytosol to acetyl CoA and oxaloacetate.

An alternative mechanism for transporting acetyl groups across the mitochondrial membrane involves the synthesis of acetylcarnitine:

$$\text{Acetyl CoA} + \text{Carnitine} \rightleftharpoons \text{Acetylcarnitine} + \text{CoA.}$$

Acetyl carnitine is transported from the mitochondrion to the cytosol, where a similar enzyme catalyses the formation of acetyl CoA and carnitine (compare the reactions for fatty acyl carnitine on p. 204A).

A

Biosynthesis of fatty acids

Fatty acid synthesis proceeds by addition of two-carbon units to acetyl CoA; malonyl CoA is the intermediate by which these units are added. The coenzyme NADPH carries out the reductive steps.

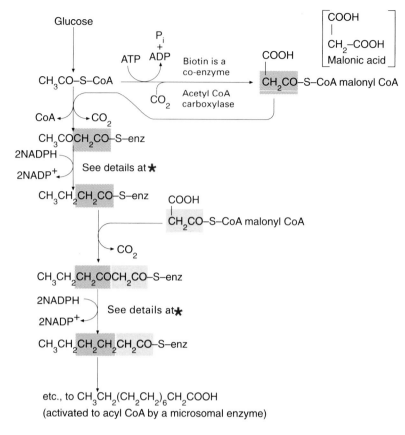

★ Details of the reductive steps:

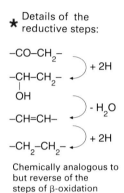

Chemically analogous to but reverse of the steps of β-oxidation

---

B

Palmitic acid synthesized by fatty acid synthase is converted to palmitoyl CoA in the endoplasmic reticulum.

Palmitic acid + ATP + CoA ⟶ Palmitoyl CoA + ADP + P_i

C

Regulation of acetyl CoA carboxylase.

Regulation of fatty acid synthesis can be exerted at the point of synthesis of malonyl CoA by *acetyl-CoA carboxylase*.

The effect of malonyl CoA on carnitine acyltransferase helps to prevent oxidation of newly synthesized fatty acid (see p. 204).

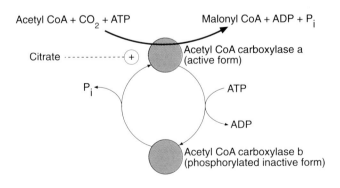

Acetyl-CoA carboxylase is strongly activated by citrate, and is subject to control by phosphorylation to an inactive form by a protein kinase and conversion to the active form by a phosphatase. Biotin is required as coenzyme, CO_2 being introduced via carboxybiotin (see p. 136B).

Fatty acid structures

Palmitic acid synthesized by fatty acid synthase is only one of a number of different fatty acids required by the body. *Palmitoyl CoA* can be elongated by microsomal enzyme systems, which add carbons as two-carbon units, utilizing acetyl CoA, to the carboxyl end of the molecule. Different desaturase enzymes can insert double bonds at specific carbons between C-1 and C-9. Thus there is Δ^4-desaturase, Δ^5-desaturase and Δ^9-desaturase.

There are no mammalian desaturases which insert double bonds between C-9 and the methyl group. Thus, as the body needs fatty acids with double bonds in this region, precursors known as essential fatty acids must be taken in from the diet. Linoleic and α-linolenic acids are examples.

Examples of fatty acids are shown below.

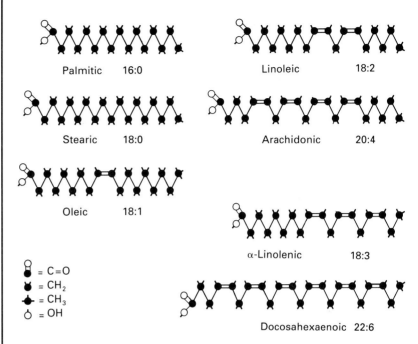

Palmitic   16:0

Stearic   18:0

Oleic   18:1

Linoleic   18:2

Arachidonic   20:4

α-Linolenic   18:3

Docosahexaenoic 22:6

$\mathbf{Q} \atop \bullet$ = C=O
$\bullet$ = CH$_2$
$\blacktriangleleft$ = CH$_3$
$\circ$ = OH

Linoleic acid cannot be synthesized by the body and must be provided in the diet and is hence termed an essential fatty acid. Linoleic acid serves as a precursor for arachidonic acid. Oleic acid and stearic acid can be synthesized from palmitic acid and thus need not be provided in the diet.

As indicated above, a fatty acid can be designated by the convention (No. of carbons):(No. of double bonds). Thus, linoleic acid, with 18 carbons and two double bonds, is 18:2. The double bond position can be designated in two ways:

1. From the carboxyl group, when the symbol Δ is used. Thus, linoleic acid is $\Delta^{9,12}$
2. From the methyl end, when the symbol n- (or ω-) is used. Thus linoleic acid is n-6,9 (or ω-6,9).

The advantage of the second convention is that it emphasizes a metabolic relationship between the acids that depends on the fact that, in mammals, double bonds cannot be introduced enzymically on fatty acid carbons nearer to the methyl end of the molecule than the ω-9 carbon.

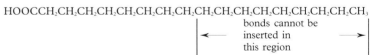

HOOCCH$_2$CH$_2$CH$_2$CH$_2$CH$_2$CH$_2$CH$_2$CH$_2$CH$_2$CH$_2$CH$_2$CH$_2$CH$_2$CH$_2$CH$_2$CH$_2$CH$_3$

bonds cannot be
$\longleftarrow$ inserted in $\longrightarrow$
this region

This means that there are 'families' of fatty acids, arising from precursor acids taken in the diet; these are shown on the next page.

A

There are three major families of fatty acids, derived from precursors with ω-6 or ω-3 double bonds or from palmitic acid.

ω-6 family

Linoleic acid

$(18:2 \; \omega\text{-}6,9) \xrightarrow[\text{desaturation}]{\text{Elongation +}}$ γ-Linolenic acid $\longrightarrow$ Arachidonic acid
$\qquad\qquad\qquad\qquad\qquad\qquad$ (18:3 ω-6,9,12) $\qquad$ (20:4 ω-6,9,12,15)

ω-3 family

α-Linolenic acid $\xrightarrow[\text{desaturation}]{\text{Elongation +}}$ Docosahexaenoic acid
(18:3 ω-3,6,9) $\qquad\qquad\qquad\qquad$ (22:6 ω-3,6,9,12,15,18)

ω-9 family

As double bonds can be inserted at ω-9–10, and points nearer to the carboxyl group than this, these acids can by synthesized from the saturated fatty acid palmitate, which in turn can be synthesized from acetyl CoA:

Palmitic acid $\longrightarrow$ Stearic acid $\longrightarrow$ Oleic acid $\longrightarrow$ Eicosatrienoic acid
$\qquad\qquad\qquad\qquad\qquad\qquad\qquad\qquad\qquad\qquad\qquad\qquad\qquad$ (20:3 ω-9,12,15)

Eicosatrienoic acid is synthesized in animals when no ω-6 acids are contained in the diet, and it substitutes, to some extent, for the long chain unsaturated fatty acids such as arachidonic acid that are then much reduced in the membrane lipids. Absence of the ω-6 acids in the diet is known as essential fatty acid deficiency and may be corrected by the addition to the diet of ω-6 acids such as linoleic acid.

B

Formation of triacylglycerols

Triacylglycerol synthesis involves sequential addition of fatty acyl CoA molecules to glycerol 3-phosphate to form phosphatidic acid, which is hydrolysed to diacylglycerol. This is then acylated to triacylglycerol.

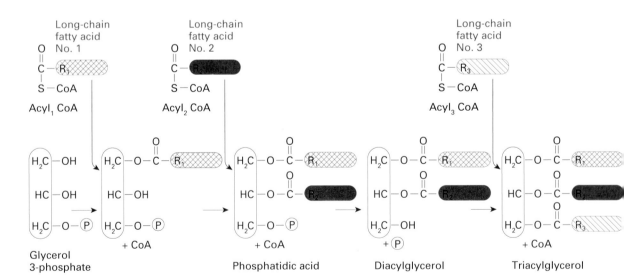

A

The importance of triacylglycerol.

Triacylglycerol is stored in a variety of storage sites (fat deposits or 'depots'). These range from the major subcutaneous adipose tissue sites, and major internal sites, to smaller storage sites within muscle tissue. It is synthesized mainly in liver and adipose tissue, but other tissues have a variable capacity to synthesize triacylglycerol.

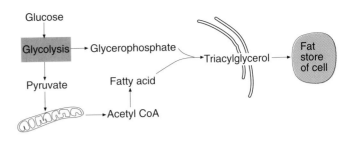

Fatty acid must be converted to fatty acyl CoA before esterification of the fatty acyl group to glycerophosphate to form triacylglycerol. The enzymes for these reactions are located in the endoplasmic reticulum.

In muscle, there are fat stores which can provide fatty acids for use as an energy reserve during exercise. The muscle of domestic animals, as a result of selective breeding and intensive feeding, normally contains fat reserves far in excess of the animal's needs of an energy reserve. Thus, so-called 'lean' meat, even after the removal of all gross visible fat, contains large amounts of triacylglycerol. The muscle of wild animals contains much smaller amounts of fat, even when they are kept in captivity.

B

The secretion of triacylglycerol as a component of lipoproteins.

Triacylglycerol synthesized in liver is exported as a major component of very low-density lipoprotein (VLDL), as described on the following pages. It is then transported in the blood to peripheral tissues such as adipose and muscle. These take up the fatty acid, after its release by lipoprotein lipase.

Apolipoprotein B_{100} (see next page) is synthesized in the rough endoplasmic reticulum and takes up lipid synthesized in the smooth endoplasmic reticulum (especially triacylglycerol) as it is transported to the Golgi complex, with eventual formation of mature VLDL. Secretory vesicles are formed and transport the VLDL to the plasma membrane, where it is released into the bloodstream *via* the Space of Disse and sinusoids.

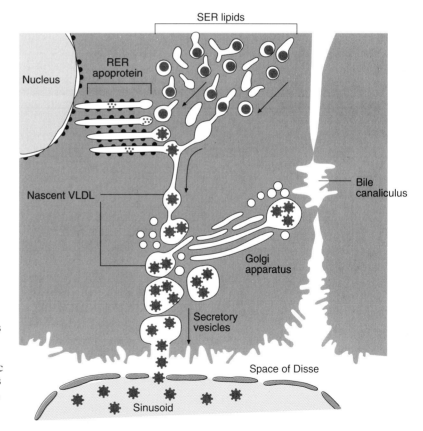

A

Chemistry and metabolism of plasma lipoproteins

Lipids circulate in the blood as complex molecules called *lipoproteins*.

Lipoproteins are large assemblies of lipid and protein. As the different classes of lipoprotein vary in their relative content of lipid they have a wide range of densities. They are also of widely different sizes. They can be separated either by electrophoresis, or by centrifugation in solutions of appropriate density. This has led to their classification as shown below.

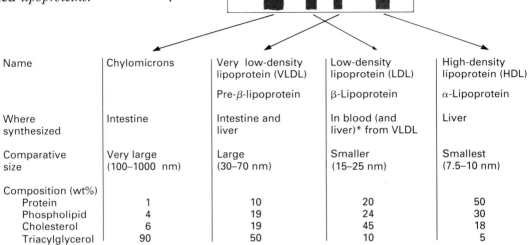

Paper electrophoresis

Name	Chylomicrons	Very low-density lipoprotein (VLDL)	Low-density lipoprotein (LDL)	High-density lipoprotein (HDL)
		Pre-β-lipoprotein	β-Lipoprotein	α-Lipoprotein
Where synthesized	Intestine	Intestine and liver	In blood (and liver)* from VLDL	Liver
Comparative size	Very large (100–1000 nm)	Large (30–70 nm)	Smaller (15–25 nm)	Smallest (7.5–10 nm)
Composition (wt%)				
Protein	1	10	20	50
Phospholipid	4	19	24	30
Cholesterol	6	19	45	18
Triacylglycerol	90	50	10	5

*   It is not certain whether all β-lipoprotein is formed from VLDL in blood or whether some may be directly synthesized in liver.

B

The composition of the lipoproteins can be represented in diagrammatic form.

The proteins associated with lipoproteins have been purified and characterized and are termed apolipoproteins (or simply Apo B, etc.).

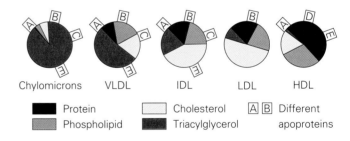

Chylomicrons   VLDL   IDL   LDL   HDL

- ■ Protein
- ▦ Phospholipid
- ☐ Cholesterol
- ▨ Triacylglycerol
- Ⓐ Ⓑ Different apoproteins

The different apoproteins are designated A_I, A_{II}, B, C_I, C_{II}, C_{III}, D and E. Each lipoprotein has a characteristic pattern of apoproteins.

Chylomicrons and VLDL	LDL	HDL
A_I, B, C_I, C_{II}, C_{III}E	B	A_I, A_{II}, DE

Apo B synthesized by liver has M_r 514 000 and is designated B-100 to distinguish it from that synthesized by intestine (B-48) which is approximately 48% of the size of B-100 (see p. 102D for protein synthesis).

IDL = Intermediate Density Lipoprotein, an intermediate stage between VLDL and LDL (see page 223A).

A

Lipoprotein lipase

An important enzyme in plasma is lipoprotein lipase, which hydrolyses triacylglycerols and reduces *chylomicrons* and VLDL to smaller fragments. This enzyme used to be called *'clearing factor'*, as after a fatty meal it cleared the milky look of plasma caused by large quantities of chylomicrons being absorbed from the intestine.

Lipoprotein lipase: hydrolyses triacylglycerol in chylomicrons and VLDL to fatty acids and glycerol.

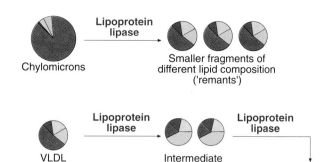

Lipoprotein lipase is activated by apoprotein C_{II}.

B

In order for fatty acids of triacylglycerols to enter cells, the triacylglycerol must be hydrolysed at the plasma membrane. The fatty acids can then be transported through the membrane into the cell.

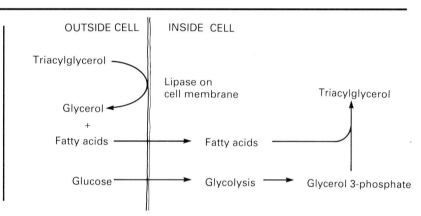

In the cell, triacylglycerol may be resynthesized from the fatty acids, but glucose is essential to provide glycerol 3-phosphate via the glycolytic pathway. The glycerol released outside the membrane can be oxidized to CO_2 and H_2O in liver, or can form glucose, after phosphorylation to glycerol 3-phosphate by glycerol kinase.

C

Enzyme action in the intestine.

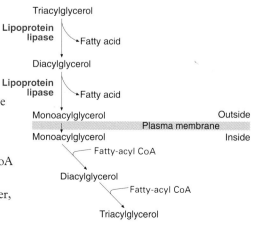

In the intestine, monoacylglycerol and fatty acids result from the action of lipase. The interstitial cells of the intestine contain enzymes that can catalyse the reaction

Monoacylglycerol + 2 Fatty-acyl CoA
$\longrightarrow$ Triacylglycerol + 2 CoA

This forms an alternative pathway for triacylglycerol synthesis, independent of a direct supply of glycerol 3-phosphate. However, an oxidizable substrate is needed, as energy is required for the synthesis of the fatty-acyl CoA derivatives from the fatty acids released by the lipase.

A

Formation of cholesterol esters

HDL is thought to be synthesized as small discs, rich in unesterified cholesterol and lecithin. These discs are converted to HDL as a result of the action of lecithin–cholesterol acyltransferase, which enriches the lipoprotein in esterified cholesterol.

Action of lipoprotein lipase on VLDL causes formation of a molecule less rich in triacylglycerol (IDL). IDL accepts cholesterol esters from HDL, to form LDL.

Note that lecithin is an older name for phosphatidylcholine (see page 236).

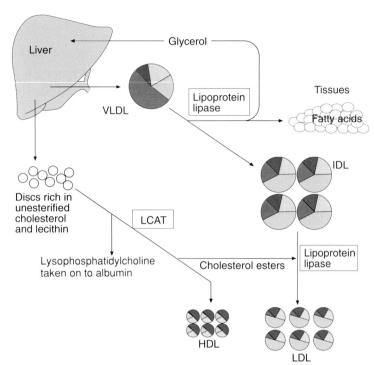

The cholesterol of lipoproteins is predominantly in the esterified form. These esters are synthesized from cholesterol and lecithin by the enzyme lecithin–cholesterol acyltransferase (LCAT, pronounced 'el cat'), found in plasma. The enzyme is activated by apoprotein A_I.

$$\text{Phosphatidylcholine} + \text{Cholesterol} \underset{\xleftarrow{\hspace{1cm}}}{\overset{\text{LCAT}}{\rightleftharpoons}}$$
$$\text{Lysophosphatidylcholine} + \text{Cholesterol ester}$$

B

Apolipoproteins

The apoprotein moieties of the plasma lipoproteins have a variety of functions.

Apoprotein	Function
A_I	This apoprotein is an activator of lecithin–cholesterol acyl transferase
B-100	Apoprotein B-100 is recognized by receptors on liver cells and other cells within the peripheral circulatory system, and plays an important role in the uptake by these cells of lipoproteins that carry this apoprotein
C_{II}	Apoprotein C_{II} activates lipoprotein lipase. A deficiency of this apoprotein has in some cases been associated with elevated plasma triacylglycerol levels
E	Liver cells carry receptors for apoprotein E, which is important for efficient uptake by liver of lipoproteins in which it occurs

An overview of lipoprotein metabolism

Fat is absorbed from the intestine as chylomicrons, synthesized in the intestinal cells. LDLs are synthesized from VLDL by enzymes in the plasma (some VLDLs are synthesized in the intestine).

LDL *distributes* cholesterol to tissues
HDL *removes* cholesterol from tissues

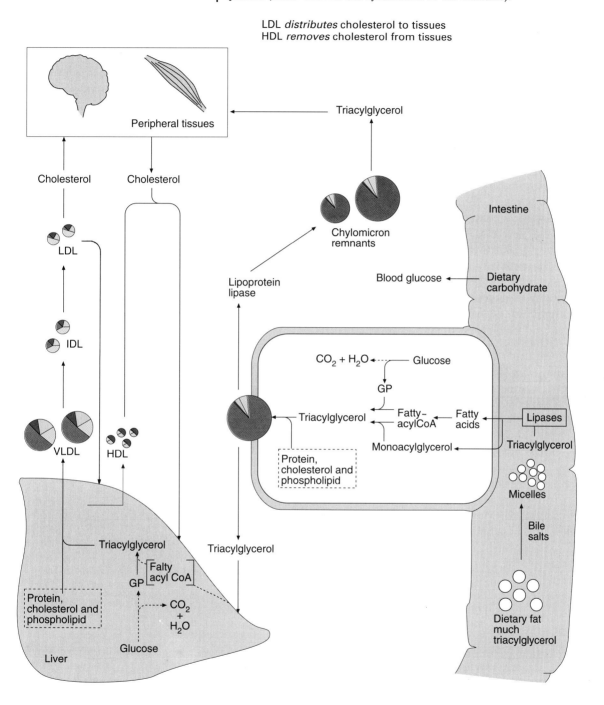

A

Biosynthesis of cholesterol

Cholesterol is one of the major constituents of plasma lipoproteins.

Cholesterol synthesis in liver is the subject of much interest because of the evidence that high blood cholesterol levels may be associated with increased risk of *atherosclerosis*. *3-Hydroxy-3-methylglutaryl-CoA reductase* is a key enzyme that regulates the activity of the pathway for cholesterol synthesis located in the smooth surfaced endoplasmic reticulum.

3-Hydroxy-3-methylglutaryl-CoA reductase (HMG-CoA reductase) is an important key enzyme that regulates cholesterol synthesis.

B

Further metabolism of *isopentenyl diphosphate* (a five-carbon compound) results in polymerization of six five-carbon compounds to squalene (30 carbons), which is converted in a further series of reactions to cholesterol.

Cholesterol is an important precursor of the sex hormones, the steroid hormones of the adrenal cortex, and of the the bile acids. All cells may take up cholesterol from the blood as it is a major constituent of the plasma lipoproteins. The major site of cholesterol synthesis is the liver, from which it is secreted as a lipoprotein constituent. Many tissues can, however, synthesize cholesterol, including the adrenals and reproductive organs, and also skin fibroblasts, which can be readily cultured and used to investigate cholesterol metabolism in humans.

Formation of bile acids and the excretion of cholesterol

The only way that excess levels of cholesterol can be removed from the body is by conversion to bile acids and excretion in the bile.

The bile acids are synthesized from cholesterol. Two distinct series of acids are synthesized, one yielding *cholic acid,* and the other *chenodeoxycholic acid.* From these, *deoxycholic acid* and *lithocholic acid* can be synthesized in the gut by bacteria. The acids are secreted into the bile largely as *glycocholic* and *taurocholic acids.*

In cholic acid, all three hydroxyl groups project from the same face of the molecule (see p. 228C). The synthetic reactions are thus directed to achieving this, including inversion of the 3β-hydroxyl group in cholesterol to a 3α-hydroxyl group in cholic acid.

Taurocholic acid is a compound in which taurine is linked to cholic acid by an amide linkage between the carboxyl group of cholic acid and the amino group of taurine. *Taurine* ($H_2NCH_2CH_2SO_3H$) is formed from a derivative of cysteine, in which the sulphydryl group has been oxidized to a sulphonic acid group ($-SH \rightarrow -SO_3H$).

Glycocholic acid has glycine linked to cholic acid in a similar manner.

The enzyme that hydroxylates the 7 position (7 α-hydroxylase) regulates the rate of synthesis of the acids

Essential steps in the conversion of cholesterol to bile acids

Oxidation of side chain

Insertion of 7α-hydroxyl

Removal of double bond

Inversion of 3-hydroxyl from β to α

The 12α-hydroxyl must also be inserted in the synthesis of cholic acid

Cholic acid

Chenodeoxycholic acid

Deoxycholic acid

Lithocholic acid

A

Bile

Bile is formed in the liver and secreted down the *bile duct*, into the *gall bladder*, where it passes to the intestine. Here the bile salts facilitate the degradation of ingested fats (see p. 212B).

Bile is a complex solution of salts and protein, and contains micelles of cholesterol, phospholipids and bile salts. The composition of these micelles is critical: imbalance may result in cystallization of cholesterol in the gall bladder, leading to the formation of gallstones. This may result from small differences in composition.

Component	Normal bile (%)	Abnormal bile (taken from a patient with cholesterol gallstones) (%)
Lecithin	74	71
Bile salts	20	13
Cholesterol	6	16

B

Bile micelles include bilayers of phospholipid containing cholesterol.

These form a disc with an ionic flat surface above and below, and hydrophobic sides. These are coated with bile salts, which present their hydrophobic face to the hydrophobic region of the disc, and a hydrophilic face directed towards the aqueous medium.

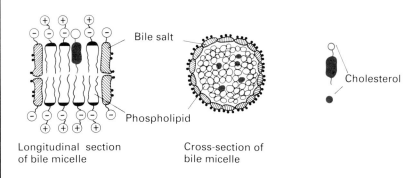

Longitudinal section of bile micelle

Cross-section of bile micelle

Bile salt

Phospholipid

Cholesterol

The bilayers of phospholipid and cholesterol are essentially similar formations to those found in cell membranes.

C

Bile acids are a highly specialized form of detergent.

However, rather than consisting of a molecule with a hydrophobic and hydrophilic *end*, as in a normal detergent, bile acids have a hydrophilic and a hydrophobic *face*.

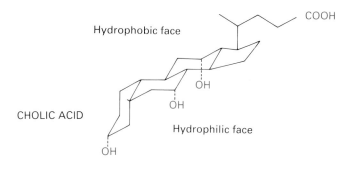

Hydrophobic face

COOH

OH

OH

CHOLIC ACID

Hydrophilic face

OH

The complex series of reactions converting cholesterol into the bile acids is directed to converting configurations of the hydroxyl groups and the ring conformations to give a structure highly specialized for its function, in which all hydroxyls project from the same face of the molecule.

A

Whole-body regulation of cholesterol metabolism

Control of the level of total body cholesterol depends on the rate of removal by conversion to bile acids in relation to the rate of synthesis in the liver. The rate of synthesis is regulated by feedback inhibition by excess cholesterol on HMG CoA reductase, so removal may only speed up synthesis. The main factor regulating cholesterol synthesis is the amount of cholesterol entering the liver as a result of uptake of chylomicrons, recycled LDL and other lipoproteins.

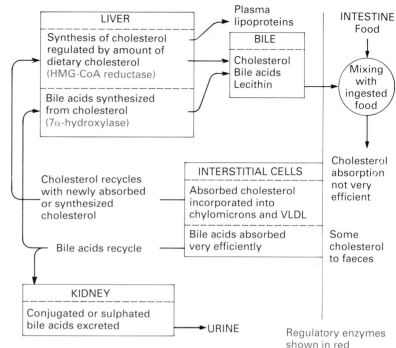

Regulatory enzymes shown in red

The bile contains mixed micelles of cholesterol, bile acids and lecithin. If these micelles contain too much cholesterol this may crystallize out and form gallstones. Excretion of cholesterol in the bile is thus concerned mainly with maintaining the composition of the bile and is not a route for controlling cholesterol excretion, although, as the absorption of cholesterol is not very efficient, some cholesterol is lost by this route. Conversion of bile acids in the gut to forms that are not re-absorbed, and urinary excretion of bile acids that have been conjugated or sulphated in the kidney, are other routes of excretion of cholesterol.

B

Control of the level of plasma LDL

In the normal individual, LDL (carrying the B-apoprotein) and IDL the lipoprotein from which it is formed (carrying B- and E-apoproteins), are removed from the plasma after binding to apoprotein receptors on the surface of liver and other cells.

There is continual recycling of LDL back from the plasma into the liver, mediated by the Apo B receptor (see next page). The activity of HMG-CoA reductase is regulated by the amount of cholesterol entering the liver. The rate of synthesis of the receptor, and thus the number of receptors is down-regulated by the rate of uptake of LDL.

Much IDL is removed from the circulation before it is converted into LDL. The E-apoprotein is lost before conversion of IDL to LDL. The rate of removal of IDL from the blood regulates to some extent the rate of its conversion to LDL.

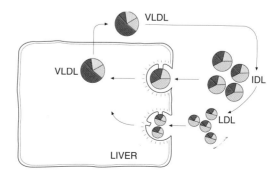

LDL receptor traffic

In normal individuals, LDL has only a limited life in the plasma, after which it is taken up by cells and degraded. It binds through its B-100 apoprotein to a receptor in a *clathrin-coated pit* in the plasma membrane. It is internalized as a *coated-vesicle*, and separated from its receptor at the endosome stage (see p. 275). The receptor is recycled to the plasma membrane, whilst the vesicle containing the LDL fuses with a lysosome. The cholesterol esters of the LDL are hydrolysed to cholesterol and fatty acid, and the phospholipids, triacylglycerols and apoprotein are degraded. Free cholesterol that is internalized inhibits the synthesis of the Apo B receptor, and also inhibits HMG-CoA reductase, reducing synthesis of cholesterol *de novo*. Elevated cholesterol levels also activate the synthesis of storage cholesterol esters by acyl-CoA:cholesterol acyltransferase (ACAT).

NOTE: Apo E receptors are also found on plasma membranes, especially of liver, and are thought to regulate the uptake of Apo E-bearing lipoproteins.

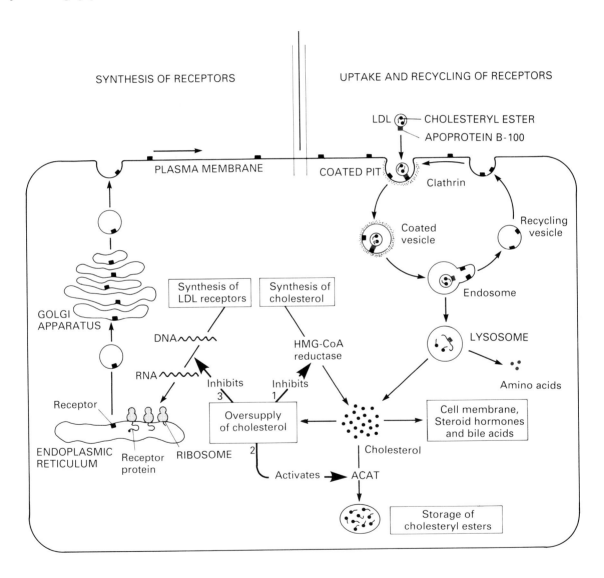

A

Hyperlipidaemias (hyperlipoproteinaemias)

Patients in whom high plasma lipoprotein levels are found may often be classified into types, first described by Fredrickson. In some cases (comparatively rarely) there is a familial trait that gives rise to the abnormality.

Types of hyperlipidaemia

Fredrickson type	Lipoprotein elevated	Cholesterol level	Triacylglycerol level
I	Chylomicrons (also possibly VLDL)	+	+ + +
IIa	LDL	+ + +	±
IIb	LDL and VLDL	+ +	+ +
III	'Floating' LDL	+ + +	+ + +
IV	VLDL	±	+ +
V	VLDL and chylomicrons	+	+ + +

±, normal to slightly increased; + +, moderately increased; + + +, greatly increased.

B

Electrophoresis of plasma proteins, using a stain for lipid, may be used to aid diagnosis of *hyperlipoproteinaemias*.

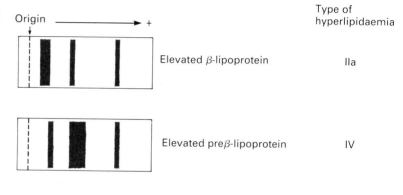

By comparison of the electrophoresis patterns with the composition of individual lipoproteins on page 222A it can be seen how the elevated plasma level of cholesterol or triacylglycerol arises in each type of hyperlipidaemia.

C

A more extensive analysis of the lipoprotein pattern is revealed by the analytical ultracentrifuge.

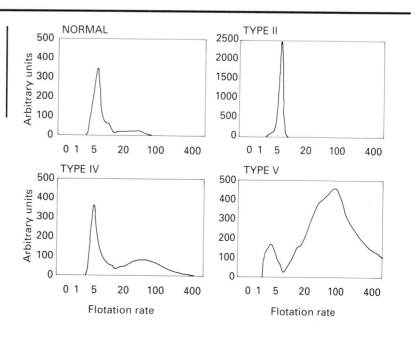

The scale on the abscissa relates to the density of the lipoprotein and is marked in Svedberg flotation units (S_f). Note the different scale on the ordinate for type II.

LDL comprises the S_f 0–20 fractions, VLDL S_f 20 upwards.

A

Type II familial hyperlipoproteinaemia.

The genetic basis of this condition is known.

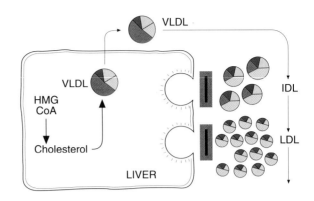

In the homozygote, the gene for the synthesis of the Apo B receptor is totally lacking. This leads to exceptionally high levels of LDL (and thus also of cholesterol) in the blood. In the heterozygote the gene of only one chromosome is functional, leading to the synthesis of reduced concentrations of receptor, and to increased risk of early death due to atheromatous plaques in the arterial wall. In the homozygote, death may occur in childhood, in heterozygotes often before the age of 40 years.

B

Uptake by macrophages of excess 'aged' LDL.

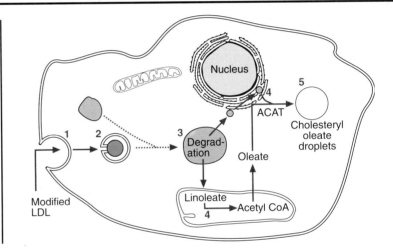

A receptor in the plasma membrane of macrophages is capable to taking up 'modified LDL'. In the absence of normal LDL receptors (as in type II hyperlipidaemia) high levels of LDL circulate in the plasma for extended periods, and may become modified in such a way that they can be taken up by this macrophage receptor. Over a long period of time these macrophages become engorged with lipid and can be seen in histological sections as 'foam cells' deposited in arterial walls. They store esterified cholesterol in lipid droplets, but with oleoyl as the acyl moiety rather than linoleoyl, which predominates in plasma.

1. 'Modified LDL' receptor
2. Internalization of 'modified LDL' in vesicles
3. Degradation in lysosomal particle: protein to amino acids, phospholipids and triacylglycerol to fatty acids and glycerol phosphate or glycerol, and cholesteryl linoleate to cholesterol and linoleate
4. Cholesterol enters endoplasmic reticulum (ER); linoleate degraded to acetyl CoA in mitochondria; acetyl CoA converted to oleoyl CoA (via palmitate) in cytosol and ER; formation in ER of cholesteryl oleate by ACAT
5. Accumulation of cholesteryl oleate in lipid droplets.

A

Cholesterol and bile acid recycling

Unless converted to bile acid, cholesterol recycles continuously round the body under the action of the plasma lipoproteins. Extensive reabsorption of bile acids exerts feedback inhibition on the synthetic pathway, limiting the excretion of cholesterol by this route. Reduction of body cholesterol is thus dependent on the fraction of bile acid that escapes to the faeces. Poor absorption of cholesterol allows some ingested cholesterol to pass into the faeces.

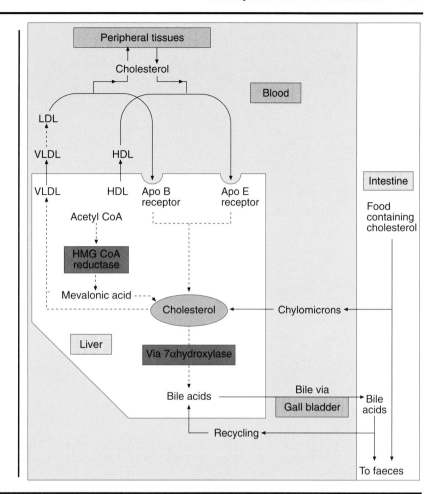

B

Therapeutic strategies for lowering cholesterol

Bile acid excretion is utilized in developing strategies for the treatment of type II hyperlipidaemia. However, this alone does not achieve adequate lowering of the plasma cholesterol level, and it is also necessary to use an inhibitor of cholesterol synthesis.

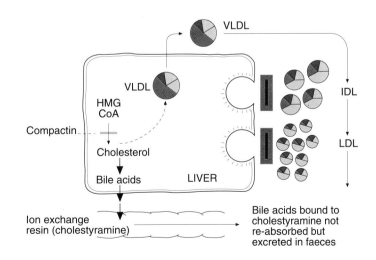

A two-pronged approach to therapy is required. An anion exchange resin (cholestyramine), taken by mouth, binds bile acids in the gut and prevents their reabsorption. However, any effect that this might have in lowering plasma cholesterol would be counteracted by increased synthesis of cholesterol in the liver as a result of the lowering of negative feedback on cholesterol synthesis. An inhibitor of HMG-CoA reductase, such as *compactin* or *mevinolin* (lovastatin), can be administered in addition to prevent this.

Dietary polyunsaturated fatty acids and atherosclerosis

The type of fatty acid in ingested fat influences the composition and to some extent the levels of plasma lipoproteins.

The fatty acid profile of the diet is reflected in the fatty acid pattern of plasma lipoproteins. Increasing the amount of linoleate in the diet will increase the linoleate level in all the lipoprotein types, and thereby the polyunsaturated:saturated (P/S) ratio. For reasons not fully understood, this causes a lowering of the plasma cholesterol level, and official recommendations are to increase the dietary P/S ratio to 2. The difficulty of achieving this can be appreciated by examining the fatty acid content of different meats and oils, some of which are listed in the table. It can be seen that poultry and fish have high levels of polyunsaturated fatty acids, and these are favoured in dietary recommendations for reducing the risk of heart disease. In fish, the fatty acids are of the long-chain ω-3 family, and ingestion of fish oil tends to lower triacylglycerol levels rather than cholesterol (again, the reasons are not well understood). Not all oils used for cooking are equally rich in polyunsaturated fatty acids, but oil is generally preferable to solid fat (lard or beef fat), which is highly saturated.

Fat or oil	16:0	16:1	18:0	18:1	18:2*	20:4*	20:5[†]	22:6[†]
Animal fat								
Beef	35	3	20	36				
Mutton	27		31	32				
Pork	30	3	11	42	11			
Rabbit	27	9	3	28	20			
Chicken	18	10	5	34	17			
Duck	21	6	6	49	16			
Turkey	23	7	8	32	25			
Fish								
Herring	15	12	2	21	3	1	9	6
					(20:1 ω-9,20	and	22:1 ω-11,30)	
Sardine	16	9	3	11	1	2	17	13
Salmon	11	5	4	24	5	5	5	17
Oils								
Olive	10		2	78	7			
Corn	10		2	30	50			
Sunflower	6		6	18	69			

Compositions vary considerably from sample to sample. Only selected fatty acids are listed but values represent % of all fatty acids present.

*ω-6.

[†]ω-3.

8

CARBOHYDRATE AND FAT METABOLISM

D. Phospho-lipids, other lipid substances and complex carbohydrates

Increasingly numerous lipid and carbohydrate substances are being identified that have essential roles in a variety of cell and tissue functions. For the most part, these are membrane associated. Phospholipids form a two-dimensional surface, the *lipid bilayer* of the plasma membrane and other membranes, into which proteins are inserted. The plasma membrane also contains *glycolipids*. *Cholesterol*, despite a poor reputation gained from its association with heart disease, is nevertheless an essential constituent of the plasma membrane. Complex carbohydrates and *glycolipids* are constituents of the cell surface. They are also important in connective tissue and in the matrix that surrounds cells, the *glycocalyx*, or in tissues, the *basal lamina*. There is a dynamic interaction between these molecules, concerned with the behaviour of the cell in relation to its environment. Their function is dealt with more fully in Chapter 9. This present chapter is primarily concerned with their structures and metabolism.

Structure of phospholipids

Membrane lipids consist of:
1. Phospholipids
2. Sphingolipids
3. Cholesterol
4. Glycolipids (containing carbohydrate).

Phospholipids have a glycerol backbone. A fatty acid is esterified to two of the −OH groups of the glycerol. The third is esterified by a phosphate group and a nitrogenous compound (choline, ethanolamine or serine) except that one family of phospholipid contains inositol, a six-carbon sugar alcohol. The nitrogenous moiety is often referred to as a 'base', and the phosphate and 'base' together comprise the 'headgroup'.

Phospholipids isolated from natural sources vary in their fatty acids esterified at positions C-1 and C-2. Thus, many isomers (often called species) are possible for each phospholipid, i.e. 1-palmitoyl-2-linoleoyl-*sn*-glycero-3-phosphocholine and 1-stearoyl-2-arachidonoyl-*sn*-glycero-3-phosphocholine are species of phosphatidylcholine. Diacyl-*sn*-glycero-3-phosphate has the trivial name 'phosphatidic acid'. Thus, diacyl-*sn*-glycero-3-phosphocholine is often alternatively termed phosphatidylcholine.

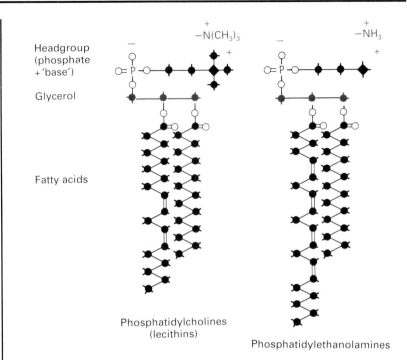

Phosphatidylcholines
(lecithins)

Phosphatidylethanolamines

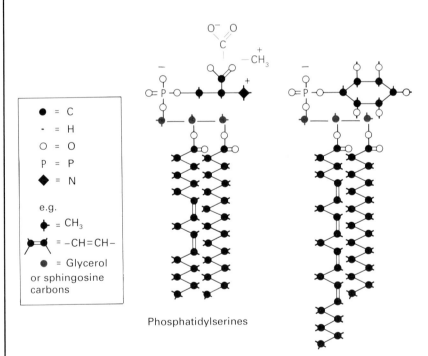

Phosphatidylserines

Phosphatidylinositols

The drawings are diagrammatic and do not attempt to represent molecular orientations in space.

A

Phospholipids are derived from *sn*-glycerol 3-phosphate.

$$
\begin{array}{ll}
CH_2OH & C\text{-}1 \\
HO \blacktriangleright C \blacktriangleleft H & C\text{-}2 \\
CH_2OPO_3H_2 & C\text{-}3
\end{array}
$$

Note: The prefix '*Sn*' indicates *s*tereospecific *n*umbering. It derives from an arbitrary convention which numbers the glycerol carbons as in the diagram, with the chiral carbon in the configuration shown.

B

Synthesis of phospholipids

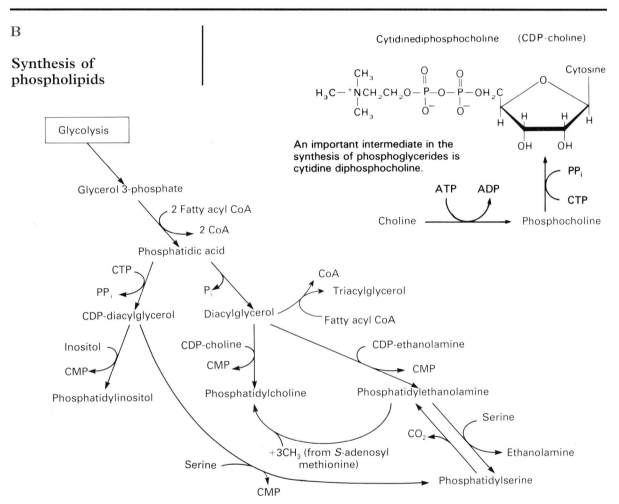

Cytidinediphosphocholine   (CDP-choline)

An important intermediate in the synthesis of phosphoglycerides is cytidine diphosphocholine.

The synthesis of phospholipids and sphingolipids involves many complex reactions and only an outline is presented here. Synthesis of phosphatidic acid involves the acylation of glycerol phosphate as previously described for the synthesis of triacylglycerol. The compounds CDP-diacylglycerol, CDP-choline and CDP-ethanolamine involve an ester linkage between the hydroxyl group of diacylglycerol, choline or ethanolamine and the terminal phosphate of CDP.

A

Phospholipases

Phospholipids play an important role in cell stimulation as substrates for phospholipases.

Phospholipase A_2 removes unsaturated fatty acids from the *sn*-2 position of phospholipids. If arachidonic acid is released, it acts as a substrate for formation of prostaglandins.

Marked in red are the bonds attacked by different phospholipases. Thus fatty acids are removed from the 1 position by phospholipase A_1 and from the 2 position by phospholipase A_2. Phospholipase C removes the whole of the headgroup, leaving diacylglycerol, and phospholipase D removes the terminal hydroxy compound, leaving phosphatidic acid.

B

Of particular importance in many mammalian cells is an enzyme with phospholipase C action which is specific for phosphatidylinositol and its phosphorylated derivatives. This enzyme has the alternative name *phosphatidylinositol phosphodiesterase*. Thus, it attacks diesters of phosphoric acid, i.e. phosphoric acid with two of its acidic −OH groups involved in ester linkage.

Release of the headgroup from phosphatidylinositol 4,5-bisphosphate yields inositol trisphosphate and diacylglycerol.

PtdIns, phosphatidylinositol; PtdInsP, phosphatidylinositol 4-phosphate; PtdInsP$_2$, phosphatidylinositol 4,5-bisphosphate; InsP$_3$, inositol 1,4,5-trisphosphate; DG, diacylgylcerol; PLC, phospholipase C.

A

Sphingolipids

The backbone of sphingolipids is the base *sphingosine* ($CH_2OHCHNH_2CHOHCH=CH(CH_2)_{12}CH_3$). A fatty acid is attached to the nitrogen in an amide linkage. To the terminal hydroxyl a sugar or chain of sugars is attached, excepting sphingomyelin, which has a phosphocholine grouping on this hydroxyl group.

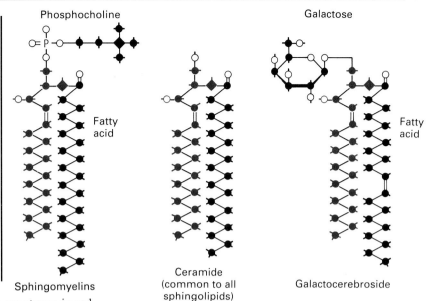

The atoms derived from sphingosine are shown in red.

The sphingolipids include sphingomyelins, cerebrosides and gangliosides. Sphingomyelin may be described as a phospholipid as it contains phosphorus. Cerebrosides and gangliosides contain no phosphorus, but do contain carbohydrate structures, and thus are both sphingolipids and glycolipids. Sphingomyelin is an important structural lipid in membrane bilayers, comprising 10–20% of the total phospholipid of the bilayer. As their name suggests, cerebrosides (and their sulphated derivatives, sulphatides) are found in appreciable quantities in the brain, especially myelin, which also applies to gangliosides. Cerebrosides and gangliosides are found in much smaller quantities in most membranes.

B

The synthesis of sphinganine.

$$\begin{array}{ccccc}
\overset{O}{\overset{\|}{C}}-S-CoA & & & CH_2OH & CH_2OH \\
(CH_2)_{14} & + \overset{CH_2OH}{\underset{COO}{H-C-NH_3^+}} & \xrightarrow{\quad H^+ \quad CO_2 \quad} & \overset{CH_2OH}{\underset{CH_3}{\overset{H-C-NH_3^+}{\underset{(CH_2)_{14}}{O=C}}}} & \xrightarrow{\quad NADPH \quad NADP^+ \quad} & \overset{CH_2OH}{\underset{CH_3}{\overset{H-C-NH_3^+}{\underset{(CH_2)_{14}}{HO-C-H}}}} \\
CH_3 & & & &
\end{array}$$

| Palmitoyl CoA | Serine | Dehydrosphinganine | Sphinganine |

C

Sphingomyelin is formed by addition of fatty acid to sphinganine, and oxidation to yield *ceramide*, which reacts with CDP-choline to yield sphingomyelin.

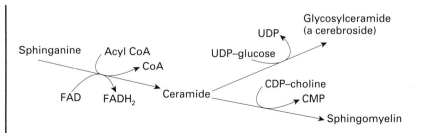

Ceramide also reacts with UDP-glucose or UDP-galactose to form glucosylceramide or galactosylceramide (the cerebrosides).

A

Steroid hormones

Cholesterol acts as a precursor to the steroid hormones, which comprise:

1. Female sex hormones, or *oestrogens*, synthesized in ovarian tissue
2. Male sex hormones, or *androgens*, synthesized in testicular tissue
3. The *glucocorticoids*, cortisol and corticosterone
4. *Mineralocorticoids*, such as aldosterone.

Progesterone itself has important hormonal properties.

The initial steps in the synthesis of all types of steroid hormone are similar, and involve the formation of progesterone. Alternative pathways sometimes proceed directly from pregnenolone through 17α-hydroxypregnenolone.

B

The metabolism of steroids is shown on the opposite page.

The glucocorticoids and mineralocorticoids are synthesized in the adrenal cortex.

Three zones of tissue can be discerned in the adrenal cortex, the zona glomerulosa, zona fasciculata and zona reticularis. Glomerulosa cells produce aldosterone, while fasciculata/reticularis cells produce mainly cortisol in the human, with some corticosterone.

Aldosterone acts on cells of the kidney distal tubules to increase reabsorption of sodium, and increases the excretion of potassium and hydrogen ions. Its secretion is regulated by the renin–angiotensin system. Angiotensin I is formed from its precursor, angiotensinogen, by the action of the enzyme renin which is secreted by the kidney in response to osmotic changes. Angiotensin II is formed from angiotensin I by converting enzyme. Angiotensin II stimulates production of aldosterone by the adrenal cortex, and both compounds have effects that raise the blood pressure. An inhibitor of the converting enzyme, Captopril, is a widely used drug for the control of hypertension.

The main androgen, testosterone, is produced in Leydig cells of the testis.

Androgens are responsible for the development, maintenance and function of the male reproductive organs and male secondary sex characteristics, such as the ability to form and deliver sperm, low pitch of the male voice and the pattern of hair distribution. They have profound anabolic activity, increasing muscle and bone mass, an action utilized by farmers and at times (illegally) by athletes.

Oestrogens, together with progesterone, are produced by cells of the Graafian follicle in the ovary.

Oestrogens are responsible for the development, maintenance and function of reproductive organs, sexual activity cycles and secondary sex characteristics in the female.

The adrenal cortex can also produce precursors in the synthesis of sex hormones, including dehydro*epi*androsterone and androstenedione. Both compounds are androgenic.

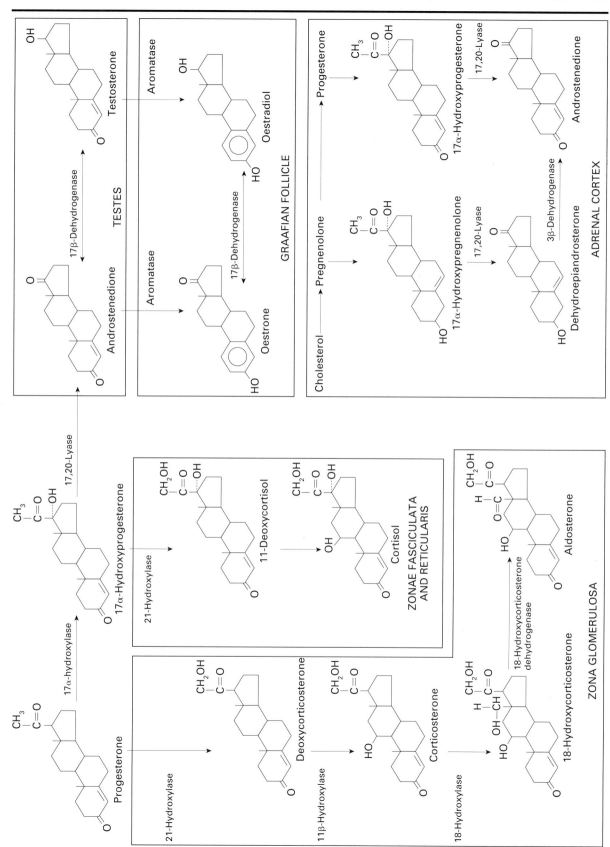

A

Feedback regulation of cortisol synthesis by the adrenal cortex results from cortisol action on the pituitary.

Inherited diseases are known, which involve a deficiency of one or other of the enzymes of the synthetic pathways shown on the previous page.

Cortisol is the major glucocorticoid and increases the blood sugar level by promoting reactions leading to gluconeogenesis (see p. 196). It also has an important anti-inflammatory action, mimicked by drugs such as cortisol and betamethasone.

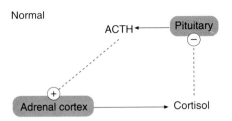

Normal

Inhibition by cortisol of ACTH secretion by the pituitary reduces the plasma ACTH level, and thus reduces the stimulus to the adrenal cortex to synthesize cortisol.

Deficiency of 21-hydroxylase also leads to failure to synthesize cortisol with resulting overproduction of ACTH (the mechanism underlying the condition known as congenital adrenal hyperplasia). In the case of 21-hydroxylase deficiency, there is not only failure to synthesize aldosterone and cortisol, which requires this enzyme, but overproduction of 17-hydroxyprogesterone and androstenedione due to the action of excessive concentrations of ACTH on the diseased adrenal.

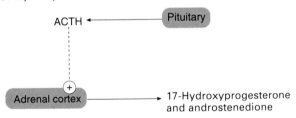

21-Hydroxylase deficiency

B

Structure of betamethasone.

A

The lipid-soluble vitamins A, E, K and D

Vitamin A

At least three different physiological functions are dependent on proper vitamin A nutrition:

1. *Somatic function*— including growth and differentiation, e.g. of epithelial structures and bone
2. *Reproduction*—it is essential for spermatogenesis, oogenesis, placental development and embryonic growth
3. *Visual process*—for vision in the dark.

Retinal

11-*cis*-Retinal

Vitamin A (retinol) has a similar structure to retinal but with a primary alcohol in place of the aldehyde group.

Vitamin A is important for night vision: 11-*cis*-retinal interacts with the protein opsin to form the visual pigment rhodopsin:

$$\text{11-}cis\text{-Retinal} + \text{Opsin} \rightarrow \text{Rhodopsin}$$

If rhodopsin is exposed to light, the double bond at C-11 of *cis*-retinal isomerizes from *cis* to *trans*, causing dissociation of rhodopsin to opsin and the *trans* form of retinal, which is regenerated enzymically to 11-*cis*-retinal. Vitamin A deficiency causes a number of other symptoms, but the mechanism of its action in preventing these is unknown. Excessive doses of vitamin A are highly toxic.

It is also now suspected that excessive doses of vitamin A could be teratogenic, a fact which is of considerable significance as vitamin A and its derivatives are an effective treatment for acne.

B

Vitamin E

The main function of vitamin E appears to be to act as an antioxidant, preventing the formation of free radicals in cell membranes. (A free radical can be represented as R•.)

Vitamin E
(α-tocopherol)

Vitamin E is an antioxidant probably acting to break the chain reaction that ensues when free radical formation occurs (see p. 165C). Polyunsaturated fatty acids esterified in membrane phospholipids are susceptible to free radical attack, after which reaction with molecular oxygen results in formation of hydroperoxides. These oxidized fatty acids are removed by phospholipases and can then be converted to hydroxy fatty acids by glutathione peroxidase (see p. 165A). Selenium deficiency results in accelerated peroxidation of fatty acids, possibly because of a decrease in the levels of selenium-dependent glutathione peroxidase. Increased intake of vitamin E can mitigate the effects of selenium deficiency.

Free radical

Hydroperoxide

A

Vitamin K₂ (menaquinone)

Vitamin K_2 is a cofactor in post-translational formation of γ-carboxyglutamyl residues in certain proteins, especially prothrombin.

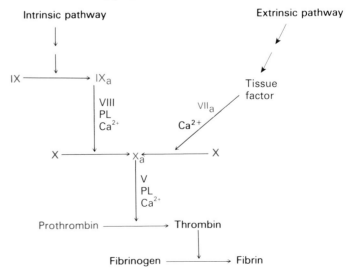

Glutamyl residue (Glu)   γ-carboxy glutamyl residue (Gla)

In vitamin K deficiency, prothrombin cannot be synthesized, and therefore the blood-clotting mechanism functions imperfectly.

B

Vitamin K exists in a reduced form and as an epoxide.

Formation of the epoxide is coupled to carboxylation of the glutamyl residue by an as yet unresolved mechanism.

Reduced form of vitamin K

Epoxide form of Vitamin K

Warfarin blocks reductase

$R =$

n varies with species

Warfarin inhibits formation of the reduced form of vitamin K that is required for the γ-carboxylation reaction, and thus mimics vitamin K deficiency. It can thus be used clinically as an anticoagulant. It is also a very effective rat poison.

C

Vitamin K-dependent carboxylation occurs in other proteins of the blood clotting cascade, and in proteins in tissues.

Simplified scheme of the coagulation mechanism with the four vitamin K-dependent clotting factors (in red). PL refers to phospholipid; V, proaccelerin; VII, proconvertin; VIII, anti-haemophilic factor A; IX, Christmas factor or antihaemophilic factor B, X, Stuart factor. The subscript 'a' means that the factor is in the active form.

Carboxyglutamate-containing proteins are powerful Ca^{2+}-binding proteins, because of the chelating properties of the γ-carboxyglutamyl group.

Intrinsic pathway   Extrinsic pathway

IX ⟶ IX_a

VIII
PL
Ca^{2+}

VII_a

Tissue factor

Ca^{2+}

X ⟶ X_a ⟵ X

V
PL
Ca^{2+}

Prothrombin ⟶ Thrombin

Fibrinogen ⟶ Fibrin

A

Vitamin D (cholecalciferol)

Vitamin D may be absorbed from the intestine or derived by the action of ultraviolet light from 7-dehydrocholesterol. It is converted by the liver to 25-hydroxycholecalciferol, which is converted by kidney to the active form of the vitamin, 1,25-dihydroxycholecalciferol.

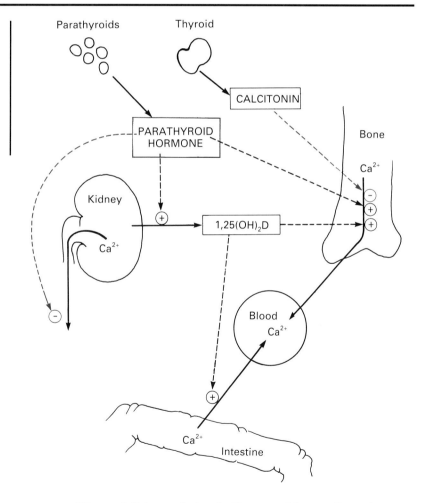

B

Calcium metabolism

The blood Ca^{2+} level is controlled by the combined action of vitamin D, *parathyroid hormone* and *calcitonin*.

Parathyroid hormone
1. is secreted by the parathyroids in response to a drop in the blood Ca^{2+} level
2. increases production of $1,25(OH)_2D$ by kidney
3. increases Ca^{2+} resorption from bone
4. decreases Ca^{2+} excretion in urine.

1,25-Dihydroxycholecalciferol
1. is the hormonally active form of vitamin D
2. increases Ca^{2+} absorption from gut
3. increases Ca^{2+} resorption from bone.

Calcitonin
1. is produced by the chief cells of the thyroid
2. reduces Ca^{2+} resorption from bone.

Dietary deficiency of vitamin D causes rickets, a disease in which there is defective bone formation. Calcitonin may be administered in the treatment of Paget's disease.

Prostaglandins

The prostaglandins are synthesized in many tissues in minute amounts. The different prostaglandins have highly potent pharmacological actions that include contraction of smooth muscle, including that of the uterus, vasodilation and platelet aggregation.

Prostaglandins are synthesized from the two endoperoxides PGG$_2$ and PGH$_2$, which are the products of the action of the enzyme *cyclooxygenase* on arachidonic acid.

In recent years it has been found that another enzyme, *lipoxygenase*, acts on arachidonic acid to form a series of hydroxy fatty acids, one of which is termed 12-hydroxyeicosatetraenoic acid (HETE). A number of pharmacological actions of these acids are being identified (they are chemotactic, for example, for macrophages, and thus may be important in the inflammatory response).

The different prostaglandin classes (see opposite page) are distinguished by the substituents on the ring part of the structure, and the different series by the number of double bonds (this depends on which acid is utilized as the precursor of the prostaglandin).

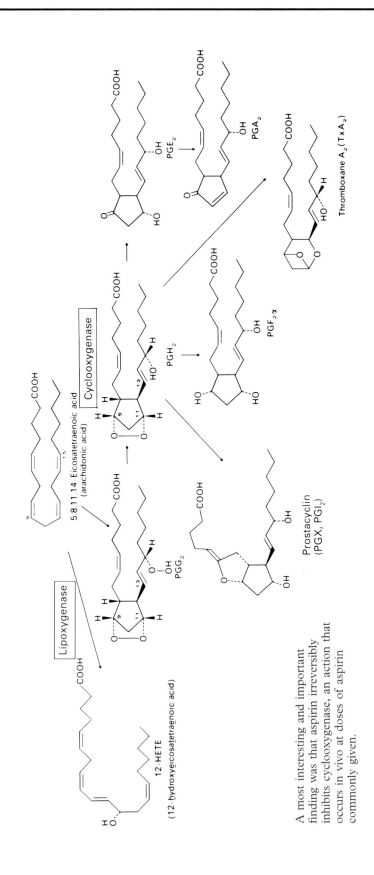

A most interesting and important finding was that aspirin irreversibly inhibits cyclooxygenase, an action that occurs in vivo at doses of aspirin commonly given.

Structure of prostaglandins and related compounds

Common features	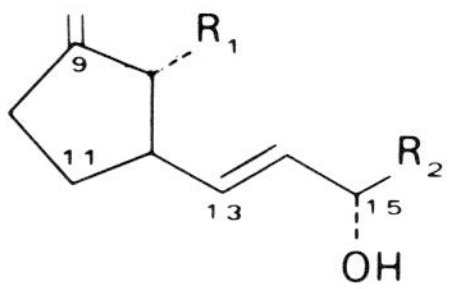
Different classes	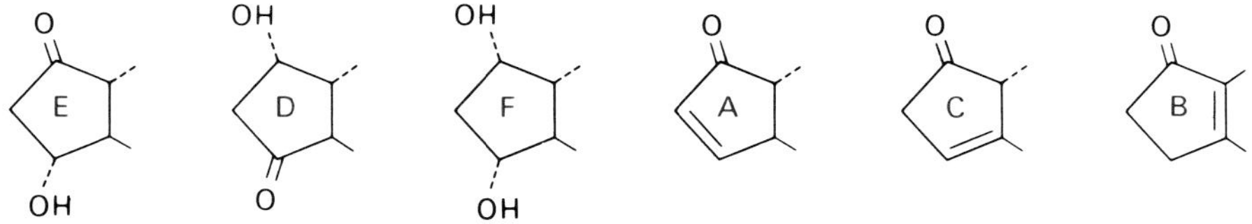
Different series	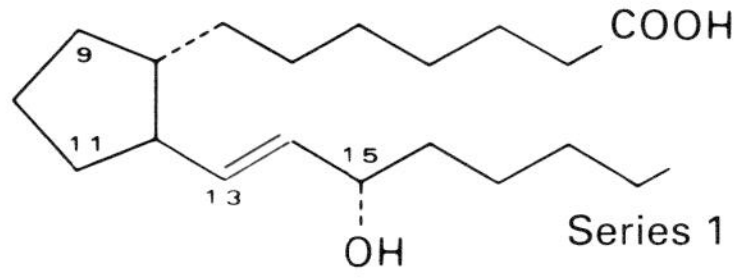 Series 1 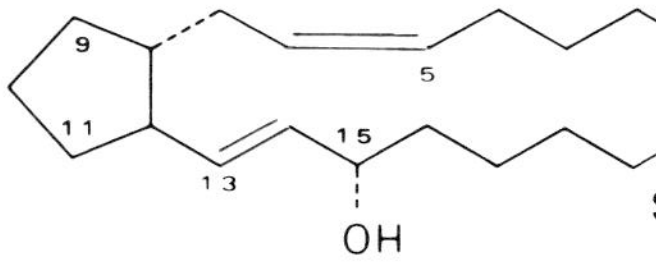 Series 2 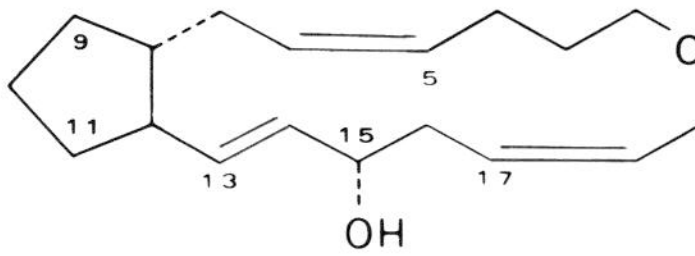 Series 3
Endoperoxides	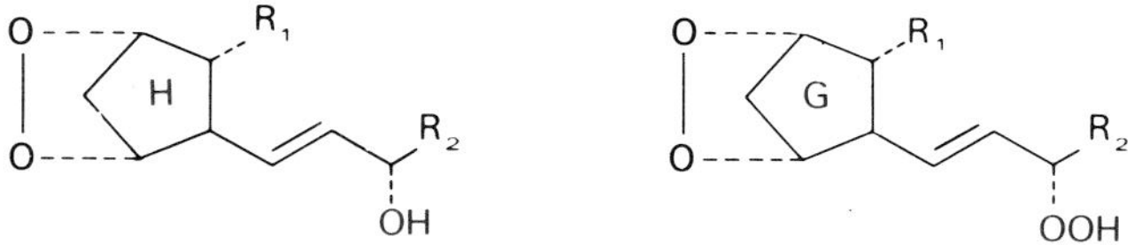
Thromboxanes (Txs)	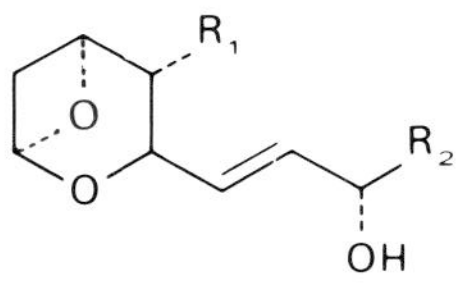 TxA 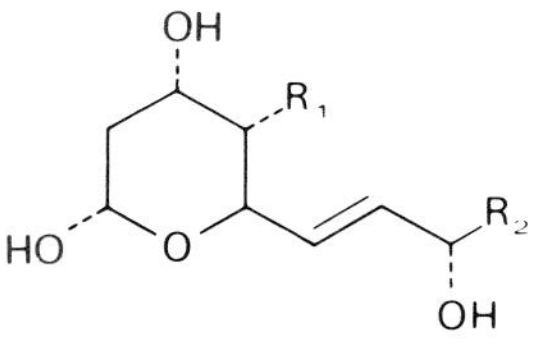 TxB
Prostacyclin	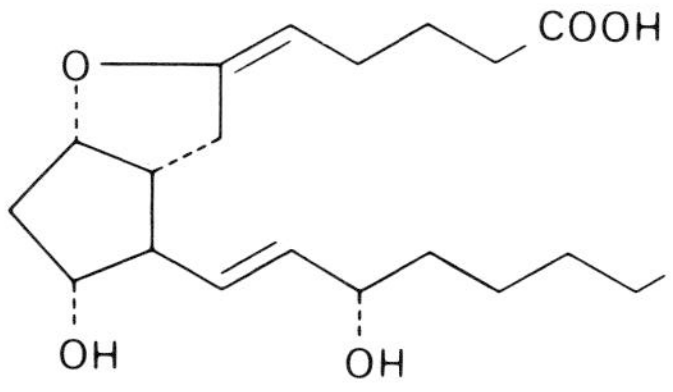

A

Linoleic acid can be elongated and desaturated to dihomo-γ-linolenic acid, which gives rise to the '1' series of prostaglandins. Metabolism of α-linolenic acid gives rise to the '3' series of prostaglandins.

Linoleic acid	Elongation	$C_{20:3}$	Oxygenation	
(18:2 ω-6,9)	Desaturation	ω-6,9,12		→ PGE_1
α-Linolenic acid	Elongation	$C_{20:5}$	Oxygenation	
(18:3 ω-3,6,9)	Desaturation	ω-3,6,9,12,15		→ PGE_3

The most commonly found series is the '2' series, synthesized from arachidonic acid.

B

Leukotrienes

The *leukotrienes* are arachidonic acid metabolites.

When conjugated to glutathione these are known as the peptidoleukotrienes. However, many polyunsaturated fatty acids undergo oxidation by different lipoxygenase enzymes to form hydroxyl derivatives, which appear to exercise a regulatory action in various cell functions. The structure of leukotriene B_4 is shown. This is chemotactic.

A recent development has been the elucidation of the structures of a series of compounds that have been termed the leukotrienes. These are derived from arachidonic acid by the action of lipoxygenases, which yield hydroxyeicosatetraenoic acids. These are conjugated to glutathione via its sulphydryl group. Slow-reacting substance of anaphylaxis (SRS-A) has been shown to be such a compound, the search for its structure leading to the delineation of the leukotriene series. This promises to be a widely distributed class of compound, with actions in the inflammatory response and related systems.

The structures of some leukotrienes are shown.

Leukotriene is abbreviated LT, thus LTC_3 denotes leukotriene C_3.

	R_1	R_2	R_3
LTC_3	Glu	Gly	C_7H_{15}
LTC_4	Glu	Gly	C_7H_{13}
LTD_3	H	Gly	C_7H_{15}
LTD_4	H	Gly	C_7H_{13}

LTC_4 and LTD_4 derive from arachidonic acid, whilst LTC_3 and LTD_3 derive from $C_{20:3}$ ω-9,12,15 which lacks the double bond in this region.

Leukotriene B_4

A

Structures of complex carbohydrates

Complex carbohydrates (*oligosaccharides*) contain a wide variety of monosaccharides, including those that are modified. Modification includes acetylation, phosphorylation and sulphation and are often bound to protein, to give the glycoproteins. They may also be bound to *ceramide* to give *gangliosides* and other complex glycolipids. These are often found at the outer surfaces of cells.

Submaxillary mucin (porcine A blood-group specificity)

$$GalNAc(\alpha1\rightarrow3)Gal(\beta1\rightarrow3)GalNAc\text{-O-Ser}\left\{or\ Thr\right\}$$

$$\underset{\substack{\alpha1 \\ Fuc}}{\overset{2}{\uparrow}}\qquad \underset{\substack{\alpha2 \\ NeuNAc}}{\overset{6}{\uparrow}}$$

GalNAc = N-acetylgalactosamine   Fuc = Fucose   Glc = Glucose
Gal = Galactose   NeuNAc = Sialic acid   Cer = Ceramide

Gal, galactose; GalNAc, *N*-acetylgalactosamine; Fuc, fucose; NeuNAc, sialic acid; Glc, glucose.

The linkages between the sugars are shown in the parentheses. The blood group substances are an example of the complex carbohydrates. The antigenic activity resides in the carbohydrate moiety, and is exhibited in both the glycoprotein and glycolipid forms of the substance.

In some cases, the carbohydrate chain is linked to protein via the amide nitrogen of asparagine, *N*-linked glycoprotein. In other cases, the carbohydrate chain is linked to protein via the –OH group of serine or threonine, *O*-linked glycoprotein.

B

Fucose and *sialic acid* are important monosaccharides in complex carbohydrate structures.

α-L-Fucose

N-Acetylneuraminic acid (sialic acid)

C

A typical and important ganglioside is termed G_{M1}. Other gangliosides are formed by the addition of further sialic acid residues (NeuNAc) to this structure.

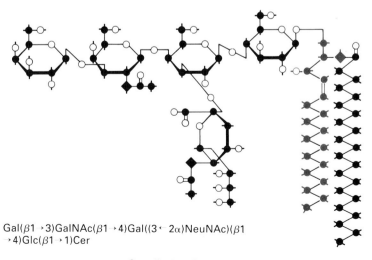

Gal($\beta1\rightarrow3$)GalNAc($\beta1\rightarrow4$)Gal(($3\leftarrow2\alpha$)NeuNAc)($\beta1\rightarrow4$)Glc($\beta1\rightarrow1$)Cer

Ganglioside G_{M1}

A

Glycoproteins

Many proteins carry carbohydrate residues, usually in oligosaccharide chains attached to an asparagine (*N*-glycosylation) or to a serine or threonine (*O*-glycosylation)

The monosaccharide unit linked to asparagine is always β-*N*-acetylglucosamine.

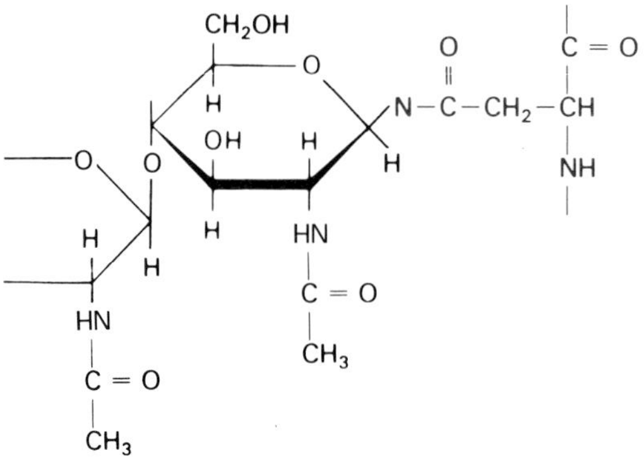

B

Intercarbohydrate links.

Two molecules of similar six-carbon sugars can be linked in many different ways, depending on whether the link between them is α-(1 → 4), α-(1 → 3), β-(1 → 4) and so on. Three molecules thus can be linked in an even greater diversity of ways. When the monosaccharides themselves differ, the number of possible variations increases.

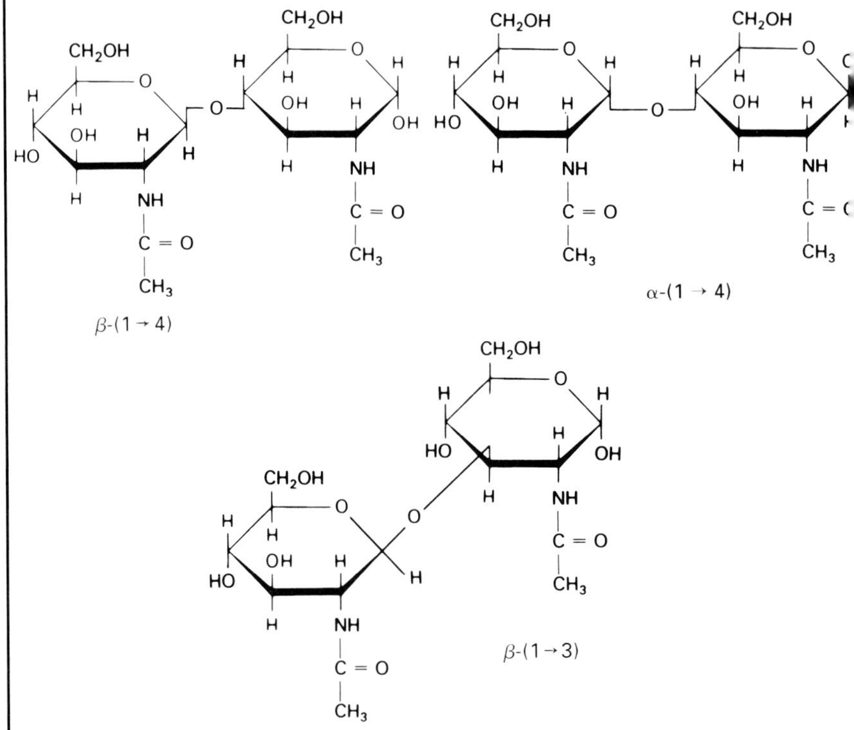

β-(1 → 4)

α-(1 → 4)

β-(1 → 3)

The three-dimensional structure will be profoundly affected by the type of linkage between molecules. Thus, the number of shapes that can be made by even a trisaccharide is very great, and one role that the carbohydrates on proteins may play is that of recognition, for which they are ideally suited due to the great number of different identities they can assume.

A

The synthesis of complex polysaccharides involves *dolichol phosphate*, a lipid molecule.

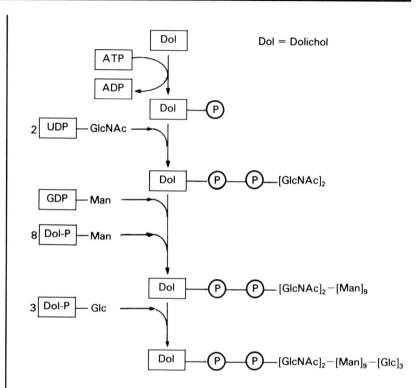

Dolichol phosphate

B

Glycoprotein biosynthesis

N-glycosylation.

In the initial steps of synthesis, the oligosaccharide chain is built up on dolichol, embedded in the membrane of the endoplasmic reticulum, and involves nucleotide-linked or dolichol-linked monosaccharides. The oligosaccharide chain is then (see C below) transferred to the amido group of asparagine in a protein, after transfer from the endoplasmic reticulum to the Golgi complex, where further carbohydrate processing takes place.

Dol, dolichol.

C

Morphology.

The oligosaccharide is linked to the growing peptide chain in the rough-surfaced endoplasmic reticulum (rough ER).

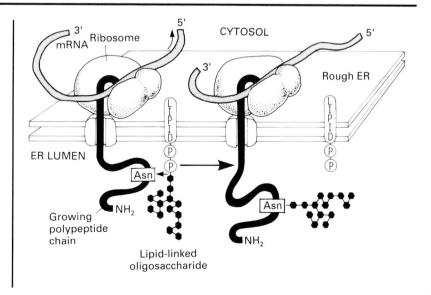

A

The core oligosaccharide.

As the nascent polypeptide passes through the Golgi complex, some of the outer units are cleaved to leave a core, shown in red. *N*-Glycosylation is inhibited specifically by *tunicamycin*, which prevents the dolichol–CHO complex being added to the asparagine.

For *O*-glycosylation see bottom of page 254.

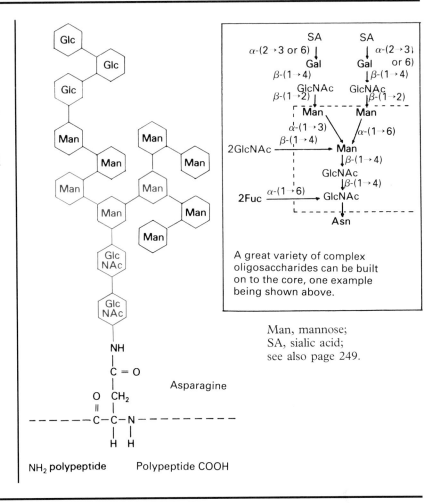

A great variety of complex oligosaccharides can be built on to the core, one example being shown above.

Man, mannose;
SA, sialic acid;
see also page 249.

B

Lactose synthesis

An enzyme that plays a key role in glycoprotein synthesis is also involved in the synthesis of lactose.

A key enzyme in the modification of the carbohydrate chains in the Golgi apparatus is *N*-acetylglucosamine galactosyltransferase. This ubiquitous enzyme is found in many types of cell bound to the membranes of the Golgi-apparatus and is a marker enzyme for this organelle (see p. 253). It adds galactose in β-(1 → 4) linkage to the terminal residues of both core oligosaccharide branches.

UDP-galactose + *N*-Acetylglucosamine $\longrightarrow$ *N*-Acetyllactosamine + UDP

Lactose is synthesized by lactose synthase which consists of two proteins, A and B. The A-protein is *N*-acetylglucosamine galactosyltransferase and the B-protein is α-lactalbumin, a milk protein. B-protein modifies the substrate specificity of A-protein from *N*-acetylglucosamine to glucose so that we have

UDP-galactose + Glucose $\xrightarrow{\text{A + B}}$ Lactose + UDP

B-protein interacts with A-protein which is bound to the membranes of the Golgi complex and this event is the initiation of lactation. The B-protein (α-lactalbumin) is excreted in the milk.

The protein secretory pathway—role of the Golgi apparatus

The Golgi apparatus is highly specialized for a variety of functions, including protein glycosylation and subsequent transport of glycoproteins to their destinations.

Three compartments have been identified in the Golgi stack, and each has a characteristic set of enzymes. These are termed the *cis*, *medial* and *trans* compartments. Each of these compartments is involved in a different stage of carbohydrate processing.

Note the complexity of the mechanisms that must exist for the budding off of vesicles and their progress to different destinations such as lysosomes, secretory storage granules or plasma membrane.

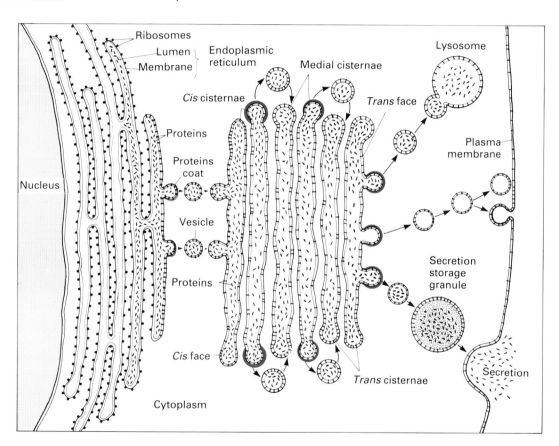

Two enzymes are present in fractions of disrupted cells containing Golgi membranes. Galactosyltransferase (see p. 252) serves as a marker for the *trans* membranes and *N*-acetylglucosaminyltransferase for the *medial* membranes. These enzymes are synthesized in the endoplasmic reticulum and are anchored in the Golgi membranes by their membrane-spanning domain.

A.

The primary structure of lysozyme and α-lactalbumin

Two proteins involved in the breakage and formation of β-(1 → 4)-glycosidic bonds are derived from a common ancestral gene.

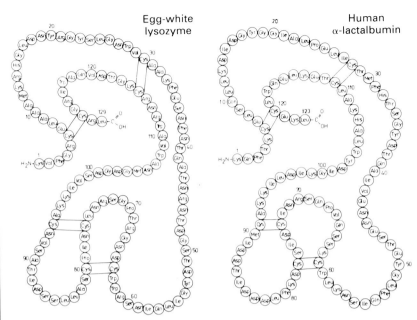

Lysozyme is a naturally occurring antibiotic, being found in virtually all tissue secretions. It hydrolyses the β-glycosidic bond between C-1 of N-acetylmuramic acid and C-4 of N-acetylglucosamine and thereby degrades the cell wall of the bacteria. Chitin, a polysaccharide in the shell of crustaceae, is also a substrate for lysozyme. Chitin consists only of N-acetylglucosamine residues joined by β-(1 → 4)-glycosidic links. As explained on page 252, α-lactalbumin plays a part in the formation of lactose, a β-(1 → 4)-disaccharide.

Above is shown the primary structure of lysozyme and α-lactalbumin. They are strikingly similar in structure in terms of both primary and tertiary structure. Also, the gene structure with respect to location of introns and exons is strikingly similar. This supports the concept that the two proteins arose by duplication of a common ancestral gene. During the course of evolution the properties of the two proteins have diverged and α-lactalbumin has no lysozyme activity nor has lysozyme activity in respect to lactose synthesis.

Many residues play a part in the action of lysozyme but particularly Glu-35 and Asp-52. In α-lactalbumin its Ca^{2+}-binding properties are important and Asp-82, 83, 87, 88 are particularly important in this respect.

B.

O-Glycosylation

O-Glycosylation is important in some glycoproteins such as cell adhesion molecules.

Addition of GalNAc to seryl or threoninyl residues occurs in the Golgi apparatus directly by reaction of UDP-GalNAc with protein (no lipid intermediate). Additional monosaccharides are then added (e.g. see p. 249) as the glycoprotein travels through the Golgi apparatus. Other forms of glycosylation through the hydroxyl group of serine or threonine are via xylose (proteoglycans, p. 146), N-acetylglucosamine in the cytosol (single monosaccharide on intracellular proteins) and fucose (EGF domains of proteins e.g. tissue plasminogen activator).

9

MEMBRANE STRUCTURE AND FUNCTION

Membranes enclose cell compartments, and the plasma membrane surrounds the cell itself, separating its contents from the environment. These membranes contain receptors signalling from the cell exterior to the interior so that the cell can make the appropriate responses to the signals it receives from hormones or growth factors, from cell–cell contacts, and from cell contacts with other surfaces. In addition, there are proteins that are concerned in transporting molecules back and forth across the membrane, including ion channels and pumps. Membranes of internal organelles are known to contain proteins involved in identifying organelles so that they can be recognized in cellular transactions, and in the plasma membrane there are cell recognition molecules identifying the cell in intercell contacts.

This chapter thus discusses: the structure of *receptors* and other surface proteins; the main *signal transduction systems* (*adenylylcyclase, G-proteins, Ca^{2+}-calmodulin* and *tyrosine kinases*); downstream *protein kinases* such as *cAMP-dependent protein kinase, protein kinase C*, and *Ca^{2+}-calmodulin-activated protein kinases; ion channels* and *pumps; clathrin* and the *endosome* system; and *cell adhesion molecules*.

A

The phospholipid bilayer

Detergents, fatty acids and similar compounds form spherical micelles when added to water.

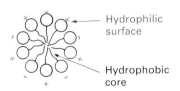

Hydrophilic surface

Hydrophobic core

The figure shows a mixture of positively and negatively charged detergents.

B

Phospholipids have a hydrophilic headgroup. The alkyl chains of the fatty acyl groups form a highly hydrophobic region.

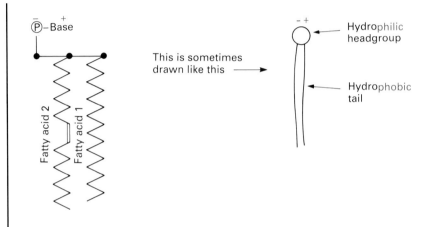

Hydrophilic headgroup

Hydrophobic tail

This is sometimes drawn like this →

Fatty acid 2   Fatty acid 1

C

Because of the geometry of the structure of phospholipids, they tend to form *bilayers* rather than micelles. These bilayers play an important role in giving membranes the two-dimensional matrix into which proteins are inserted.

Each turn of an α-helix requires 0.541 nm, so the bilayer width permits 6–7 turns.

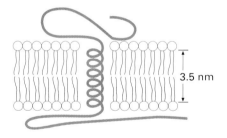

3.5 nm

Proteins that have one or more hydrophobic domains that span the membrane, usually as a helical structure, are termed *integral proteins*. The outer surface of the bilayer carries positive and negative charges that are involved in electrostatic binding of the hydrophilic domains of integral proteins that protrude on either surface of the membrane. They can also assist in binding proteins attached more loosely to the membrane surface, often referred to as *membrane-associated* proteins, although these are also bound by other mechanisms, such as by the use of a glycosyl phosphatidylinositol anchor or through fatty acylation or prenylation (see p. 266A). Not all phospholipids have strong bilayer-forming potential. Some, such as phosphatidylethanolamine, can adopt the hexagonal phase within the bilayer. This has the form of an inverted micelle, with the hydrophilic headgroups in the centre. Phosphatidylcholine strongly promotes and helps to maintain the structure of the bilayer.

A

Structure of glycophorin

The complete amino acid sequence of many membrane proteins is known, including that of glycophorin, an important protein of the red blood cell which bears some (AB and MN) of the blood group-specific carbohydrates.

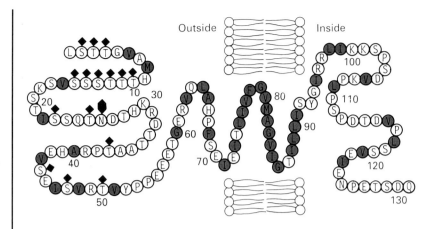

Hydrophobic side-chains (shown in red) tend to form a cluster within the bilayer. Residues 74–95 are thought to form a helix with about six turns through the core of the bilayer. Residues 62–73 may form a non-helical cluster within the membrane. Oligosaccharides, shown as squares (T/S-linked) or a hexagon (N-linked) are on the membrane outer surface. The N-terminus is residue 1.

B

Protein dynamics in the lipid bilayer

Proteins can move laterally in the plane of the membrane, and can rotate around an axis vertical to the plane of the membrane. However, they cannot 'tumble' through the plane of the membrane.

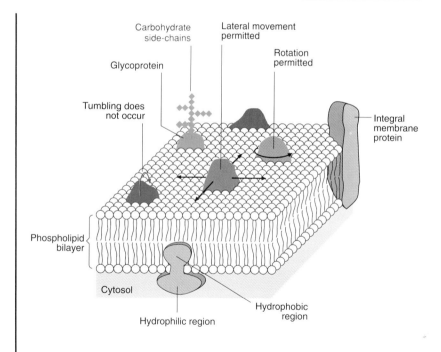

A

Receptor structure

The plasma membrane contains proteins whose primary function is to bind molecules to the external surface of the cell. Such receptors often also have the function of transmitting a signal to the cell interior (*signal transduction*) with subsequent internalization of receptor and ligand.

Receptors often have carbohydrates attached to the extracellular peptide chain.

Apo B (LDL) receptor

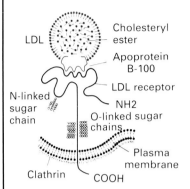

Transferrin receptor

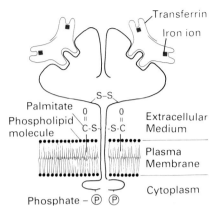

A number of the receptors that have been purified have been found to contain hydrophobic regions which anchor the receptor in the lipid bilayer. Models of this type for the LDL and transferrin receptors have been proposed.

B

Various strategies for anchoring the receptor in the lipid bilayer have evolved. Most of these involve helical domains spanning the bilayer. Orientation with respect to N- and C-terminals varies.

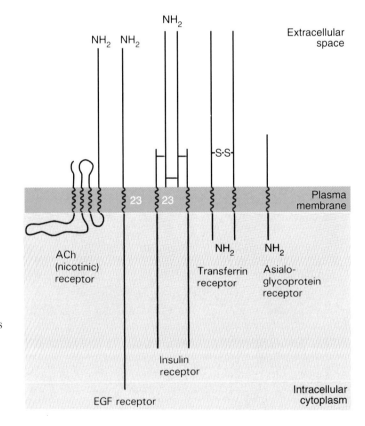

In some cases, a single peptide chain with a single pass through the membrane is involved (EGF, asialoglycoprotein). In other cases, single-pass peptide chains are linked by disulphide bonds (transferrin) sometimes *via* completely external peptides (insulin). Some receptors (nicotinic acetylcholine) pass several times through the membrane, and an important group has seven transmembrane segments (β-adrenergic—see next page).

A

The β-adrenergic receptor.

The β-adrenergic receptor traverses the lipid bilayer seven times. This is characteristic of an important group of receptors that typically act through a G-protein (see p. 264). This *heptahelical* structure gives the name 'seven transmembrane segment' (7-TMS) to this class of receptor.

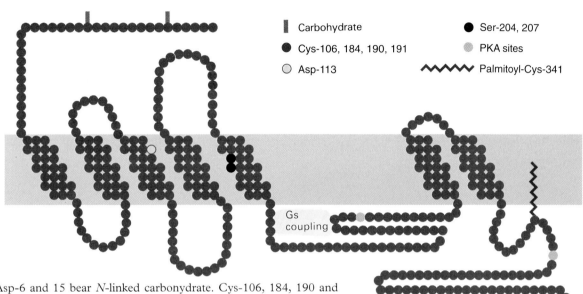

▮ Carbohydrate	● Ser-204, 207
● Cys-106, 184, 190, 191	● PKA sites
○ Asp-113	∿∿∿ Palmitoyl-Cys-341

Gs coupling

Asp-6 and 15 bear *N*-linked carbonydrate. Cys-106, 184, 190 and 191 are required for normal ligand binding and cell surface expression. As ligand binding is thought to occur in the hydrophobic region, however, these cysteines may actually stabilize the correct folded conformation. Asp-113 is conserved among all 7-TMS receptors that bind biogenic amines. Its replacement by Asn causes a great decrease in the potency of agonists in stimulating adenylylcyclase. Ser-204 and 207 are required for normal binding and activation by catecholamine agonists. A palmitoyl residue is bound to Cys-341, and penetrates the bilayer. Ser-260 and 346 are sites for phosphorylation by cAMP-dependent protein kinase (PKA).

B

The nicotinic acetylcholine receptor.

The nicotinic acetylcholine receptor has five subunits, two α, and one each of β, γ, and δ. The structure of the α-subunit has been studied in detail.

Note: The muscarinic acetylcholine receptor is of the 7-TMS class.

Sites for binding acetylcholine (ligand site) and bungarotoxin (toxin site) have been identified. As described on the following pages, the function of different residues has been explored by site-directed mutagenesis, with the results indicated; B signifies that acetylcholine binding is affected, P that gating (see p. 278) or permeation is affected, and N indicates no effect.

Outside

α   γ
β   δ

α

M4 — — M1
M3 — — M2

Inside

Toxin site

N141 B
B   C142 B
C128
Ligand site

P C192
P C193

NH₃⁺

N C222

COO⁻

4.0 nm
between
lipid
headgroups

B   P

M1 M2   M3   M4

P

N

MA

A

Fluid nature of cell membranes

Fluidity of the bilayer is evidenced by the formation of *clusters* of receptors; for example, after binding of mitogen.

The lateral mobility of proteins in the bilayer gave rise to use of the term *the fluid mosaic* membrane.

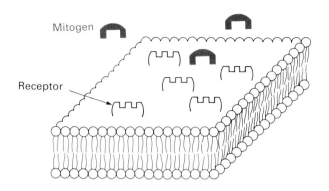

Mitogen

Receptor

Binding of mitogen first causes formation of clusters.

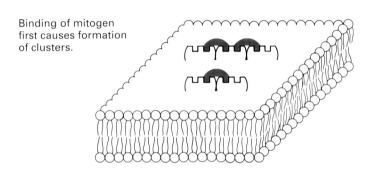

B

In certain events, such as the stimulation of lymphocytes with mitogens, clustering develops after about 30 minutes to the point of formation of a *cap* at one end of the cell. This can be seen using a light microscope if fluorescent antibodies are bound to surface antigens.

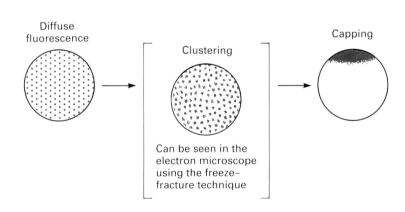

Diffuse fluorescence

Clustering

Capping

Can be seen in the electron microscope using the freeze-fracture technique

A

Cloning of the nicotinic acetylcholine receptor cDNA

An example of how protein structure and function can be probed using site-directed mutagenesis (see below)

The genes for the four subunits of the acetylcholine receptor (AChR) of *Torpedo californica* (an electric fish) have been cloned and the cDNA corresponding to each subunit was transcribed in vitro. The four mRNAs synthesized were injected into *Xenopus* oocytes (which do not normally possess any AChR). The mRNA was translated by the oocytes, which were incubated in the presence of [^{35}S]methionine.

SDS 10% PAGE of synthesized subunits

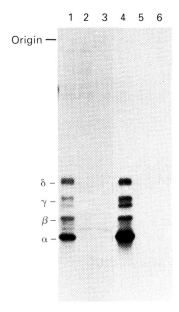

After incubation of the oocytes with [^{35}S]methionine, the AChR was precipitated by rabbit antiserum to the *T. californica* AChR and was electrophoresed in the presence of SDS on 10% polyacrylamide gel. Lane 1 is immunoprecipitate from *Xenopus* injected with *T. californica* total mRNA. Lane 4 is immunoprecipitate from *Xenopus* injected with mRNAs synthesized in vitro from cDNA. Lanes 2, 3, 5 and 6 are controls to verify that non-specific precipitation of protein by the rabbit antiserum used did not occur.

B

The membrane potential dependence of acetylcholine responses was recorded from a *Xenopus* oocyte injected with the four AChR mRNAs and then incubated for 3 days.

The oocytes that had been injected with the mRNAs and then incubated for 3 days responded to acetylcholine by causing a current, recorded under voltage clamp, at different levels of applied membrane potential, indicating that a functional AChR had been synthesized by the oocytes from the *T. californica* mRNA and incorporated into the membrane.

C

Protein engineering

The technique which allows proteins of any desired structure to be synthesized by modification of a cloned DNA is commonly referred to as *protein engineering*. The DNA can either be modified at specific sites by means of mutagens or by the replacement of stretches of nucleotides.

An example of protein engineering.

The cloning of the cDNA of the AChR has permitted site-directed mutagenesis for the introduction of deletions or specific amino acid substitutions in the subunit.

The effect of these changes of function could be tested. Each of the α-subunit-specific mRNAs synthesized with the cDNA templates with internal deletions, combined with the wild-type and subunit-specific mRNAs was injected into *Xenopus* oocytes. The oocytes were examined for the content of subunit polypeptides and for the functional properties of the AChR formed. Binding was tested by incubation with [^{125}I]bungarotoxin, which is related to acetylcholine binding, and is inhibited specifically by carbamylcholine. Acetylcholine potentials were recorded at a holding potential of −60 mV.

Subunit	Residue change* from	to	[^{125}I] Bungarotoxin binding activity	% Inhibition by carbamylcholine	Response to ACh (%)
Wild type			100	80	100
αΔ224-237	LFSFLTGL VFYLPT	PSS	0.5	NT	ND
αΔ249-257	VLLSLTV FL	SS	17	87	ND
αΔ327-334	STMKRASK	RAR	80	82	70
αΔ363-367	QTPLI	PSS	83	77	101
α128S	C	S	0.08	NT	ND
α222S	C	S	79	83	107
α141D	N	D	0.09	NT	ND
α142S	C	S	ND	NT	ND
α192S	C	S	39	27	ND
α193S	C	S	28	53	ND

*Using the single letter code for amino acids.
ND, not detectable.
NT, not tested.

Deletion mutations
The bungarotoxin-binding capacity of most of the deletion mutants was comparable with the wild type except for αΔ224–237. Segment M1 is near to the extracellular N-terminal section of the subunit, so this deletion may result in a conformational change near the binding site. Deletions in M1 or M2 resulted in loss of AChR response to acetylcholine.

Substitutions
These explored the role of cysteines which had been replaced with serines. Replacement of 128 abolished binding and AChR response, whilst 222 replacement had little effect.

The use of such techniques is now standard practice in exploring the relationship between protein structure and function.

A

Transmission at a cholinergic synapse

Acetylcholine is released into the synaptic cleft.

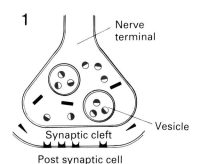

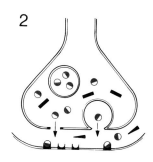

1. The nerve terminal contains acetylcholine in vesicles and in the cytosol.
2. The arrival of an action potential causes release of acetylcholine into the synaptic cleft, where it binds to the receptor in the membrane of the postsynaptic cell. (There is some discussion as to whether vesicular or cytosolic acetylcholine is released into the synaptic cleft.) Activation of the postsynaptic cell causes the action potential to be propagated in the next neurone.

B

After transmission of the action potential, acetylcholine is degraded by *acetylcholinesterase*.

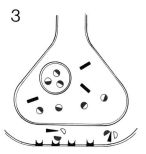

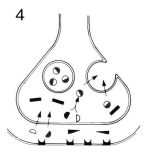

3. The enzyme acetylcholinesterase cleaves the acetylcholine to acetate and choline to prevent persistent stimulation of the postsynaptic cell.
4. Acetate and choline re-enter the presynaptic cell where acetylcholine is resynthesized by choline acetyltransferase (acetyl CoA + choline react to form acetylcholine and CoA). Endocytosis may be involved in the formation of new vesicles.

C

The nerve terminal resynthesizes acetylcholine.

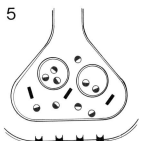

5. The nerve terminal is restored to a state in which it can await a further action potential.

⊔ Receptor in post synaptic cell membrane
▬ Choline acetyltransferase
━ Acetylcholinesterase
◔ Acetylcholine

Calcium plays an important role in the sequence. The action potential causes Ca^{2+} ions to enter the nerve terminal, and the resulting elevation of cytosolic Ca^{2+} triggers the release of the acetylcholine.

A

Adenylylcyclase and G-proteins

The action of many hormones and other agonists is mediated through activation of adenylylcyclase (see p. 67). The coupling of the hormone–receptor complex to adenylylcyclase involves a *trimeric G-protein*. Agonists which act in this way include glucagon, adrenaline, TSH and ACTH.

Some G-proteins such as EF-Tu and p21^{ras} (see next page) are monomeric having no subunits corresponding to $\beta\gamma$ (see page 105A for role of EF-Tu).

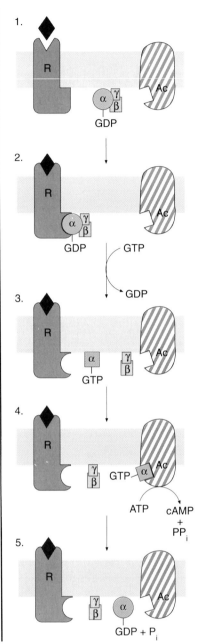

1. The receptor/adenylylcyclase system consists of a receptor (R), an enzyme (AC, adenylylcyclase) and a modulator protein, termed the G-protein (G signifies that it binds guanine nucleotides). The G-protein is trimeric, in that it consists of three subunits, α, β and γ, the α-subunit bearing the guanine-nucleotide binding site.
2. Binding of an agonist, such as a hormone (♦), to its receptor causes the G-protein to bind to the hormone–receptor complex, in which form the α-subunit can accept GTP in place of GDP.
3. After binding GTP, the α-subunit and the $\beta\gamma$-subunits dissociate from the receptor, and the α-subunit dissociates from the $\beta\gamma$-subunits.
4. GTP-α then binds to adenylyl-cyclase, activating it so that it catalyses the conversion of ATP to cAMP.
5. The α-subunit has GTPase activity, and hydrolyses GTP to GDP and P_i. This hydrolysis causes dissociation of GDP-α from adenylylcyclase, inactivating it. The G-protein trimeric state then reforms and the process can be repeated.

B

A family of G-protein α-subunits has been identified, some of which are inhibitory (α_i in the figure). The mechanism of inhibition is unknown.

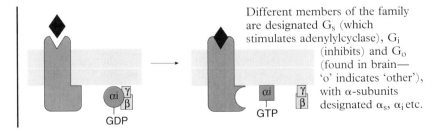

Different members of the family are designated G_s (which stimulates adenylylcyclase), G_i (inhibits) and G_o (found in brain—'o' indicates 'other'), with α-subunits designated α_s, α_i etc.

A

The G-protein superfamily, embraces molecules as diverse as the bacterial elongation factor, EF-Tu and the 21 kDa product of the *ras* oncogene (p21^{ras}). The existence of such a superfamily is a reflection of the many regulatory systems that exist using a G-protein as part of the mechanism.

The positions for the binding sites for cholera toxin (CT) and pertussis toxin (IAP, islet-activating protein, its former name, because of its effect on islets of Langerhans) are shown in the figure — see below for their action.

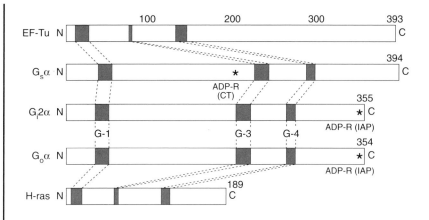

	Residues	G-1	Residues	G-3	Residues	G-4
EF-Tu	18-25	G H V D H G K T	80-84	D C P G H	135-138	N K C D
Gsα	47-54	G A G E S G K S	223-227	D V G G Q	289-295	N K Q D
Gi2α	40-47	G A G E S G K S	201-205	D V G G Q	270-273	N K K D
Goα	40-47	G A G E S G K S	201-205	D V G G Q	270-273	N K K D
Ha-ras	10-17	G A G G V G K S	57-61	D T A G Q	116-119	N K C D

In this superfamily, several highly conserved GTP-binding regions are recognized, three of which are illustrated in the figure. Much of what is known of their function derives from the X-ray structure of p21^{ras} - specific residues referred to in the following (and underlined) are those of p21^{ras}. In G-I, the amino acid sequence motif GX$_4$GK(S/T) is involved in bonding through the ϵ-amino group of Lys-16 to α- and β-phosphates of GTP or GDP. The $\underline{D}$X$_2$G motif to the G-3 region binds the catalytic Mg^{2+} through Asp-57 with an intervening water molecule. In the characteristic sequence motif of G-4, (N/T)(K/Q)X$\underline{D}$. Asp-119 hydrogen bonds to the guanine ring.

ADP-ribosylation of α$_s$ by cholera toxin prevents association of α$_s$, with βγ subunits, and thus persistently activates adenylyl cyclase. ADP-ribosylation of α$_i$ by pertussis toxin also activates the enzyme, by deactivating the inhibitory action.

B

Toxins produced by certain bacteria interact with the adenylylcyclase system by causing ADP-ribosylation. *Vibrio cholerae*, the causative agent for cholera, produces a toxin which causes ADP-ribosylation of α$_s$, and pertussis toxin (from *Bordetella pertussis*, the causative agent of whooping cough) causes ADP-ribosylation of α$_i$.

$$\text{Protein} - \overset{O}{\overset{\|}{C}} - O - 1'\,\text{Rib}\,5' - O - \overset{O^-}{\underset{\|O}{P}} - O - \overset{O^-}{\underset{\|O}{P}} - O - 5'\,\text{Rib}\,1' - \text{Ade}$$

The structure of poly(ADP-ribose) attached to a protein carboxyl group. Rib, ribose; Ade, adenine.

A

Anchoring of proteins by lipids

Fatty acyl and isoprenyl groups.

Many proteins are anchored into the plasma membrane bilayer by linkage to fatty acyl groups or isoprenoid groups which insert into the membrane bilayer. Linkage of fatty acids may be as esters or thioesters or as *N*-acyl groups. Isoprenyl groups are attached as thioethers.

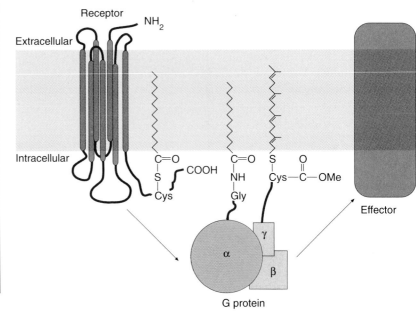

Serious inhibition of cell growth results if HMG-CoA reductase inhibitors are present in the culture medium, and this inhibition cannot be relieved by cholesterol. Investigation of this effect led to the discovery that the isoprenoids, farnesyl diphosphate and geranylgeranyl diphosphate were essential for the post-translational modification of certain proteins essential for growth. (Geranylgeranyl pyrophosphate is a 20-carbon isoprenoid synthesized from the 15-carbon farnesyl diphosphate by reaction with isopentenyl diphosphate.) It had already been found that a number of proteins undergo acylation by fatty acids such as palmitate and myristate during post-translational processing. The figure shows a G-protein-coupled seven-span receptor with palmitoylation through a thioester linkage to a conserved cysteine. The G-protein α-subunit has myristate attached in amide linkage to N-terminal glycine, and the γ-subunit is shown with a geranylgeranyl moiety attached to a C-terminal cysteine with its carboxyl methylated. The motif −CAAX (C, cysteine; A, aliphatic; X, any residue) at the C-terminal is involved in addition of the geranylgeranyl group, AAX then being removed.

B

Glycosyl phosphatidylinositol (GPI).

A number of proteins are linked to the outer surface of the cell membrane by attachment through a glycan moiety to phosphatidylinositol as shown in the structure (in this structure, the aspartate is the C-terminal residue linked to the glycan by an ethanolamine phosphate moiety).

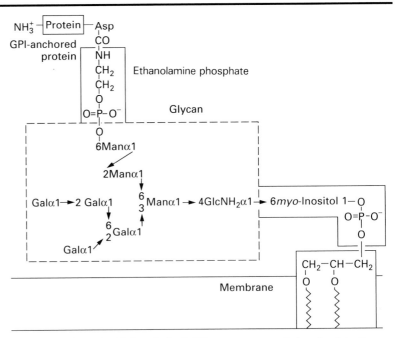

A glycosyl phosphatidylinositol of different structure is involved in the action of insulin.

A

Protein kinase A

After formation of cAMP by adenylylcyclase, the signal is transmitted by activation of cAMP-dependent protein kinase (PKA).

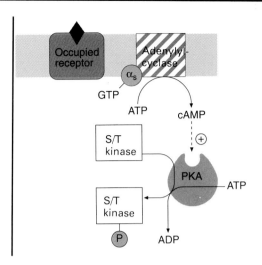

PKA is a broad-specificity protein kinase phosphorylating serine or threonine residues on a number of other serine or threonine (S/T) protein kinases (e.g. glycogen phosphorylase kinase) involved in mediating cAMP action. The characteristic target sequence is (R or K)–(R or K)–X–(S or T) e.g. –RKXS–.

B

The mechanism by which cAMP activates PKA has been shown to depend on its ability to dissociate the catalytic unit from regulatory subunits.

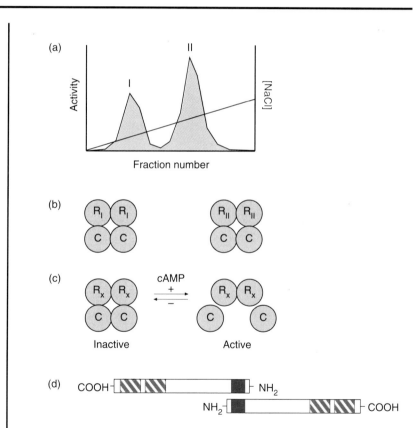

PKA can be resolved into two peaks of cAMP-dependent activity on an ion exchange column, each peak yielding a heterotetramer consisting of two regulatory subunits (R_1 from peak I and R_2 from peak II) and two catalytic (C) subunits. cAMP binds to the R-subunits, releasing active catalytic monomers.

The domain structure of the R-subunit dimer includes two cAMP-binding sites (red hatched boxes) and a dimerization domain required for R-R interactions (grey stippled boxes).

A

Protein kinase C

Another important S/T kinase is known as protein kinase C (PKC). It is activated by Ca^{2+} and diacylglycerol (DAG) and phospholipid (see also p. 270).

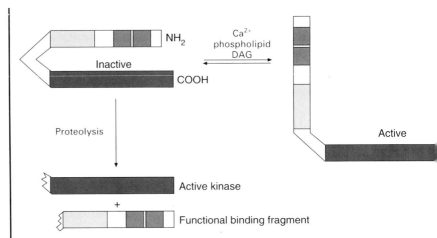

Domains identified in PKC include the kinase domain (stippled grey), two DAG/phospholipid binding sites (red) and a domain (pink) that is present in PKCs α, β and γ, but not δ, ϵ, or ζ (see below). Identification of the function of these domains has been aided by the observation that limited proteolysis splits the protein into two peptides, one of which remains catalytic, and the other binds DAG in a phospholipid-dependent manner. After partial proteolysis, the liberated catalytic domain has *constitutive* activity, that is, it is active in the absence of the activator molecules, DAG and Ca^{2+}, indicating that in the intact molecule the regulatory domain inhibits it in the absence of activators, and that DAG and Ca^{2+} act to suppress this inhibition possibly through conformational changes (as indicated).

B

The existence of a family of PKC enzymes has been revealed by cloning and sequencing techniques.

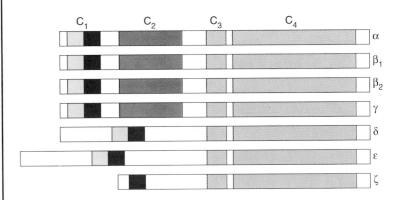

Conserved domains are designated C_1 to C_4. In all these PKCs a catalytic C-terminal domain ($C_3 + C_4$) can be defined (1) by partial proteolysis as described above and (2) by homology with the kinase domain in other types of protein kinase. PKCs δ, ϵ, and ζ do not have a C_2 domain and, as PKC ϵ (and probably δ and ζ) is Ca^{2+}-independent, it may be that the C_2 domain confers Ca^{2+} dependence. The grey and black segments represent cysteine-rich domains, at least one of which is present in all PKCs.

A

Calmodulin

Variation in the intracellular Ca^{2+} concentration is now regarded as a major regulatory system in the control of cell function. Calmodulin is an important Ca^{2+}-binding protein that mediates the action of Ca^{2+} on proteins.

Residues shown in red indicate helical regions involved in Ca^{2+} binding as discussed in B below.

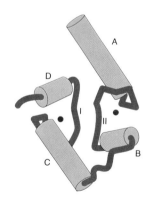

Each calmodulin molecule can bind four Ca^{2+} ions. It is a small (M_r 16 790) heat-stable protein, found in many eukaryotic cells, whose amino acid sequence has been conserved almost perfectly throughout evolution. The Ca^{2+}-dependent interaction of calmodulin with enzymes which it activates requires first the binding of Ca^{2+} with a consequent conformational change without which enzyme activation does not take place.

B

There is a relationship between calmodulin and other Ca^{2+}-binding proteins such as troponin C and parvalbumin.

A structure commonly found in Ca^{2+}-binding proteins consists of a helix–loop–helix conformation, sometimes referred to as the *EF hand*, because the E and F helices of the molecules such as troponin C in which it was first described can be envisaged as the forefinger (E) and thumb (F) of a hand in which the second finger curls back to the thumb as the intervening loop 'holding' the Ca^{2+} ion. This loop contains 6–8 oxygen atoms of Asp and Glu side-chains (see A above) which together with backbone carbonyls bind the Ca^{2+} selectively even in the presence of 10^3-fold higher concentrations of Mg^{2+}. This helix–loop–helix (HLH) domain often occurs in pairs, as is the case with calmodulin, shown in the figure, where helix A represents the 'forefinger' in this case, loop II joining it to helix B (the 'thumb'), the Ca^{2+} being represented by the small red solid circles, and a similar HLH domain comprising C–I–D. Binding of Ca^{2+} induces conformational changes that activate calmodulin to activate its target protein kinases. These tend to be rather specific S/T kinases such as myosin light-chain kinase, glycogen-phosphorylase kinase and Ca^{2+}-calmodulin (CAM) kinases I to III.

A

Phospholipase C and inositol phosphates

Phospholipase C (PLC) plays a key role in activating the Ca^{2+}-calmodulin and PKC systems described in the preceding pages. PLC hydrolyses PtdInsP$_2$ to inositol 1,4,5-trisphosphate (InsP$_3$)—see page 238.

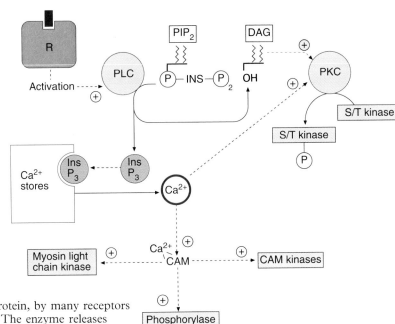

PLC is activated, possibly through a G-protein, by many receptors which have bound an appropriate ligand. The enzyme releases InsP$_3$ and diacylglycerol (DAG). InsP$_3$ reacts with receptors in the membranes of intracellular organelles which store Ca^{2+}, probably mainly in the endoplasmic reticulum. Ca^{2+} is then released from these stores and, (1) acting in concert with DAG, activates protein kinase C (PKC) and (2) activates calmodulin (CAM), which in turn activates several S/T protein kinases.

B

The inositol moiety in inositol phosphates is D-*myo*-inositol.

D-*myo*-Inositol 1-phosphate

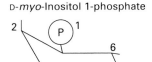

Note: the name trisphosphate is used when the phosphates are on different positions, triphosphate is used when they are linked to each other by anhydride bonds. In inositol phospholipids, a diacylglycerol group is esterified to the C-1 phosphate.

A

A considerable number of inositol phosphates can be formed from InsP_3. The products recycle to inositol, which then participates in resynthesis of PtdInsP_2.

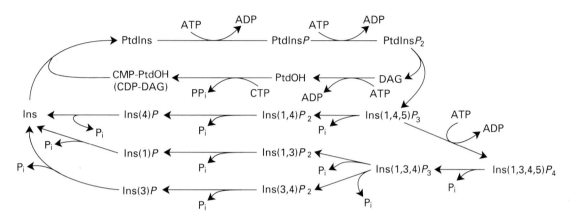

After hyrolysis of PtdInsP_2 to InsP_3, DAG is acted upon by a kinase that phosphorylates it to phosphatidate (PtdOH), which then reacts with CTP to form CMP-PtdOH. A series of phosphatases converts InsP_3 to inositol (Ins), which can then react with CMP-PtdOH to enter the pathway for resynthesis of PtdInsP_2. InsP_3 can be phosphorylated to Ins$(1,3,4,5)P_4$, which can also be converted back to Ins for resynthesis of PtdInsP_2. Few of these inositol phosphates have been assigned any role in cell physiology, and at present the reason for this complexity of inositol metabolites is not understood (a number of others not shown have been identified). In addition to the role known for InsP_3, it appears that Ins$(1,3,4,5)P_4$, acting in concert with InsP_3 brings about the entry of extracellular Ca^{2+} into the cell.

B

Retroviruses and growth factors

Much of what is known about growth regulation derives from studies of retroviruses and their associated oncogenes (see also p. 99).

When the oncogenes of retroviruses were sequenced, it was found that related genes existed in the genome of normal cells. Thus, the product of v-*sis*, an oncogene isolated from simian sarcoma virus, was found to be to a large extent homologous with platelet-derived growth factor (PDGF), a protein secreted by platelets that stimulates cell growth. Similarly, the v-*erb* oncogene (from avian erythroblastoma virus) was seen to be related to the gene coding for the receptor for epidermal growth factor (EGF), a protein that had earlier been isolated from mouse submaxillary glands and caused developmental changes when injected into new-born mice, and was also found to stimulate growth of epithelial cells.

A

Protein tyrosine kinases

One of the most important mechanisms of signal transduction involves kinases that phosphorylate protein tyrosyl residues. These are typically found in cytoplasmic domains of growth factor receptors, such as the epidermal growth factor (EGF) receptor and platelet-derived growth factor (PDGF) receptor. Another important class of protein tyrosyl kinases involves membrane-associated proteins such as the product of the proto-oncogene c-*src*.

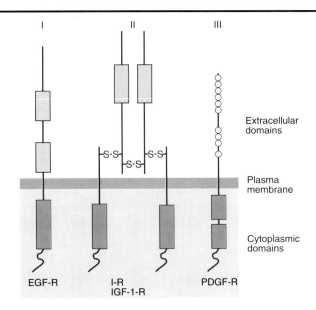

Receptor tyrosine kinases (RTKs) have been classified into three groups. Class I is typified by the EGF receptor (EGF-R), class II by the insulin receptor (I-R), and the insulin-like growth factor receptor (IGF-1-R), and class III by the PDGF receptor. The intracellular kinase domain (pink) is interrupted in class III by a long hydrophilic insertion. Classes I and II characteristically have cysteine-rich repeats (grey boxes) in the extracellular domain not present in class III receptors which do, however, have a number of conserved cysteines (open circles).

B

pp60$^{c\text{-}src}$ (indicating the 60 kDa phosphoprotein (pp) product of the cellular counterpart of the *src* oncogene from Rous sarcoma virus) and PDGF receptor (PDGF-R) have related kinase domains, except that PDGF-R has a non-kinase insert in the kinase region. These proteins undergo phosphorylation on the tyrosines indicated. pp60$^{v\text{-}src}$, the viral oncogene product, lacks Tyr-527, and mutation of this residue to a phenylalanine increases the transforming ability of pp60$^{c\text{-}src}$. *src* homology (SH) regions SH-2 and SH-3 are characteristic of proteins that associate with the tyrosine kinase class of receptors. Myristoylation (addition of myristic acid, C$_{14:0}$) is needed for membrane attachment (see p. 266A), for which a 32 kDa protein is important. The myristic acid is attached to the N-terminal glycine of pp60$^{c\text{-}src}$.

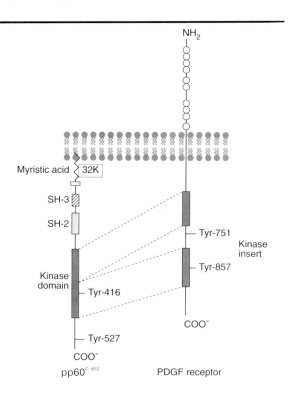

A

Oncogenes

Oncogenes provide a continuous growth stimulus by intervening in the normal growth stimulation mechanism in a non-regulated manner (see *retroviruses*—p. 271B).

Normal growth-stimulatory substances are subjected to control by regulatory mechanisms. Oncogenes differ in structure from these normal proteins in ways which allow them to escape from the normal regulatory process.

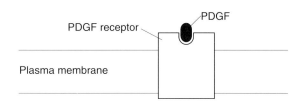

When the PDGF receptor is occupied, its protein tyrosine kinase is activated, and this stimulates cell growth.

B

There are at least three ways in which oncogene products can provide this continuous growth stimulus.

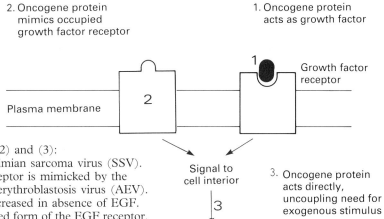

There are oncogene proteins for (1), (2) and (3):
1. An example is PDGF and v-*sis* of simian sarcoma virus (SSV).
2. Epidermal growth factor (EGF) receptor is mimicked by the v-*erb*-B oncogene product of avian erythroblastosis virus (AEV). Protein tyrosine kinase activity is increased in absence of EGF. The oncogene product is the truncated form of the EGF receptor.
3. Oncogenes of the *src* family encode proteins with tryosine kinase activity. There are other oncogenes whose products are nuclear, e.g. *myc*, *myb*, and these may play a role in activating other genes responsible for the growth response.

C

Tumour suppressor genes

Studies on retinoblastoma have revealed that development of the tumour involves loss of both alleles at a single locus, mapped to the long arm of chromosome 13 (13q14). This finding led to the concept that certain genes could be growth-suppressive. Since the initial finding with retinoblastoma, restriction fragment length polymorphism (RFLP) analysis has resulted in the mapping of the loci of inherited predisposition to a number of tumours, including (affected chromosome in parentheses) multiple endocrine neoplasia type 1 (11q), von Hippel–Lindau syndrome (3p) and neurofibromatosis type 2 (22). This predisposition is associated with the loss of alleles, involving in some cases large segments of chromosome, but with the presumption that specific genes are responsible for growth suppression when present. Loss of alleles in the germ line correlates with genetic predisposition. Recently, there has been much interest in a gene encoding a 53 kDa protein designated p53, a nuclear phosphoprotein originally discovered as an antigen in cells transformed with SV40 virus. Mutations in the p53 gene are the most common genetic alterations in human cancers. Wild-type p53 can suppress or inhibit the transformation of cells in culture by either viral or cellular oncogenes, and introduction of wild-type cDNA into a transformed cell in culture stops growth.

Growth factor receptor activation

Activation of a growth factor receptor causes phosphorylation and formation of a complex. The complex contains phospholipase C (PLC), PtdIns (PI) 3-kinase, the c-*src* 60 kDa product and *ras*-GAP, a protein that activates the GTPase activity of the *ras* oncogene product.

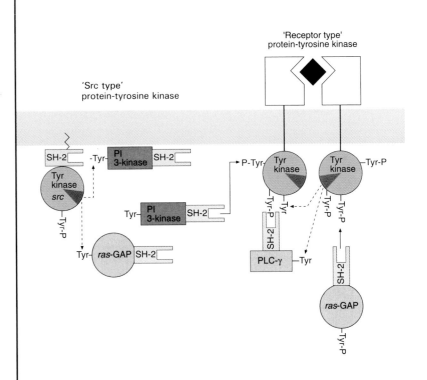

Binding of ligand to PDGF-R or EGF-R causes dimerization of the receptor, autophosphorylation of receptor tyrosines, and phosphorylation of a number of proteins, including PLC★ (see p. 270) and c-*src*. On phosphorylation, c-*src* itself then phosphorylates and activates *ras*-GAP and PI 3-kinase, a recently discovered enzyme that phosphorylates PtdIns in the 3 position of the inositol moiety (compare the enzyme that catalyses formation of PtdIns(4) *P*—see p. 271A). GAP (GTPase-activating protein) activates the GTPase of the c-*ras* oncogene product. The SH-2 domain is thought to be involved in binding these proteins to tyrosyl phosphates. Immunoprecipitation with anti-PDGF-R antibodies precipitates a complex containing these proteins, but only after activation of the receptor. Thus, it seems clear that they are involved in this signalling system, but the targets further downstream are not known. The SH-3 domain is now thought to be involved in interactions with these downstream targets.

★Several isoforms of PLC exist. One that is present in growth factor complexes is the γ-isoform (PLC-γ).

The endosome system—recycling receptors

After binding ligand, many receptors are internalized in vesicles.

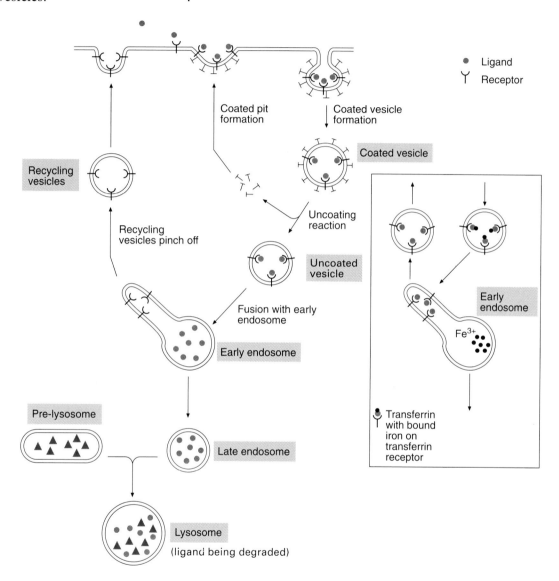

Receptors accumulate in regions of the plasma membrane referred to as *coated pits*. The inner surface of these regions is coated with the protein *clathrin*. After binding ligand, the receptors are internalized in coated vesicles. The vesicles then lose clathrin, and enter the *early endosome* system. This appears to consist of a pre-existing system of vesicles or tubules, which have the ability to segregate bound ligand and receptor, so that receptors can recycle back to the plasma membrane, allowing the liberated ligand to pass into the *late endosome* system. There are variations on this theme. For example, in the case of the transferrin receptor (see inset), transferrin and its receptor both recycle back to the plasma membrane, only the iron being dissociated. Late endosomes fuse with pre-lysosomes coming from the Golgi complex to form lysosomes, which can then degrade ligands (such as LDL —see p. 230).

A

The role of clathrin

Clathrin, the protein that controls coated pit and coated vesicle formation, exists in a conformation known as a triskelion, a trimeric, three-legged structure.

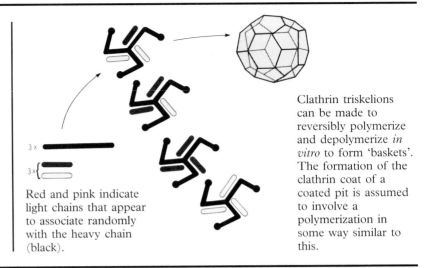

Red and pink indicate light chains that appear to associate randomly with the heavy chain (black).

Clathrin triskelions can be made to reversibly polymerize and depolymerize *in vitro* to form 'baskets'. The formation of the clathrin coat of a coated pit is assumed to involve a polymerization in some way similar to this.

B

Triskelions and isolated coated vesicles have been visualized by electron microscopy.

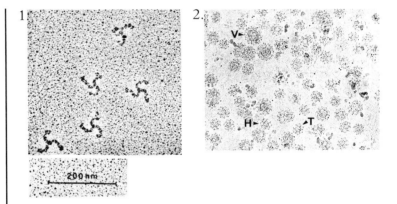

1. Individual triskelions.

2. Unstained placental coated vesicles. Hexagonal barrels (H), so-called tennis ball structures (T) and larger coats containing vesicles (V) can be seen.

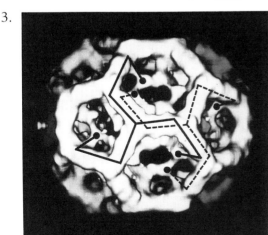

3. Three-dimensional reconstruction of a clathrin cage. This cage is a 'hexagonal barrel' which has a hexagon at the top and the bottom, six hexagons around the equator and two rings of six pentagons joining them. Two clathrin triskelions centred at the vertices between neighbouring equatorial hexagons have been superimposed diagrammatically on the reconstruction. Each leg of a triskelion runs from one vertex, along two neighbouring polygonal edges and then turns inwards with its terminal domain forming the inner shell of density seen in the reconstruction.

A

Membrane transport

Transport across cell membranes

It is obvious that communication systems must exist in the membranes to transport substances from one compartment to another, otherwise all of the compartments would be totally closed.

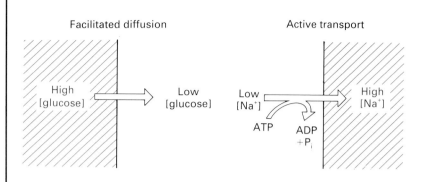

Facilitated diffusion is energy-independent and is in the direction of the concentration gradient (i.e. from high concentration to low concentration). Active transport requires the provision of energy (often involving ATP hydrolysis) and can occur against the concentration gradient.

B

Active transport

Active transport is a process whereby molecules are transported across a membrane against the concentration gradient. In some cases, this requires ATP. In other cases, a molecule can be transported against its concentration gradient by virtue of the fact that another molecule is simultaneously transported in the direction of its own concentration gradient.

Transport against the concentration gradient (i.e. from a compartment of low concentration into a compartment of higher concentration) can occur, but requires some form of energy compensation. This can be provided by hydrolysis of ATP, as in the sodium pump, that maintains sodium and potassium concentrations in nerve cells.

Action of the sodium pump

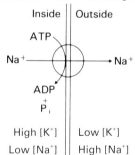

Another way of describing the action of the pump is

$$ATP \xrightarrow[Na^+ \ K^+]{Mg^{2+}} ADP + P_i$$

Isolated preparations of the enzyme hydrolyse ATP when stimulated by both Na^+ and K^+. As with other enzymes utilizing ATP, Mg^{2+} is also required.

Transport driven by an ion gradient
Glucose transport is Na^+-dependent

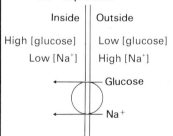

This is termed *symport*, i.e. both molecules moving in the same direction. When the molecules move in opposite directions the term *antiport* is used.

A

Components of a transport system

The components of a transport system may contain any or all of the following: a selective gate (denoted ○), a non-selective channel through the membrane, and an energy-coupling system (denoted □).

Many of the possible arrangements of these components probably occur, e.g. the gate and/or the energy-coupler may be on the inside of the membrane.

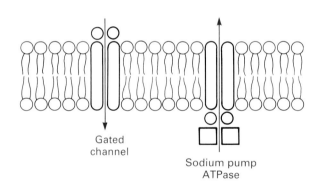

	Diffusion through bilayer	Passive transport	Gated transport	Active transport
Facilitation	−	+	+	+
Selection	−	−	+	+
Energy transformation	−	−	−	+

B

In the *nerve axon*, the *action potential* is regulated by the influx of Na^+ through a gated channel. K^+ ions then leave by other channels to return the membrane potential to normal. Ionic concentrations are eventually restored by the *sodium pump*.

Gated channel

Sodium pump ATPase

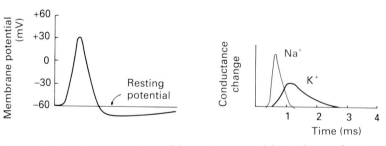

Events during the propagation of the action potential are shown above.

A

Kinetics

Transport systems show saturation kinetics, demonstrating that a transport site exists which can be occupied by only a limited number of molecules. Inhibition can occur and a K_m can be measured.

This is found both for facilitated diffusion and for active transport.

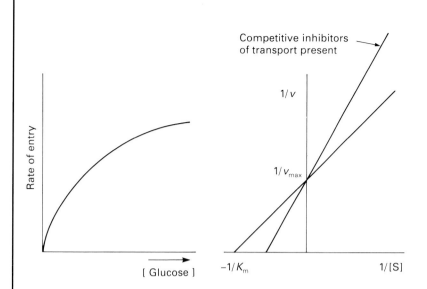

B

The red cell cytoskeleton

Beneath the cell plasma membrane there exists a system of structural proteins that are implicated in the mechanical properties of the membrane. These and other proteins form an internal system of fibres known as the *cytoskeleton*.

In the red cell, the prominent components of this system are spectrin, ankyrin, actin and tropomyosin.

In other cells actin microfilaments, microtubules and intermediate filaments are found as indicated in the following pages.

Spectrin has α- and β-chains and is thought to interact through assemblies of actin and tropomyosin. The protein ankyrin links spectrin to the plasma membrane. The band 4.1 protein (so named because of its position on gels of red cell proteins) may also be involved in linkage to the plasma membrane.

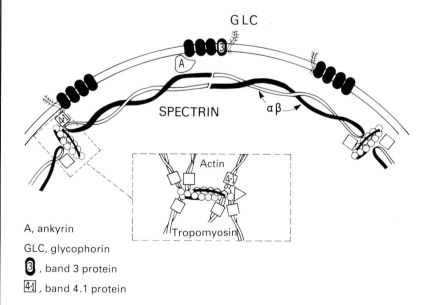

A, ankyrin

GLC, glycophorin

, band 3 protein

, band 4.1 protein

A

Actin microfilaments

The cytoskeleton can be visualized by freeze–etch electron microscopy after treatment of the cell to remove most other proteins. In this electron micrograph of a fibroblast actin filaments are so visualized.

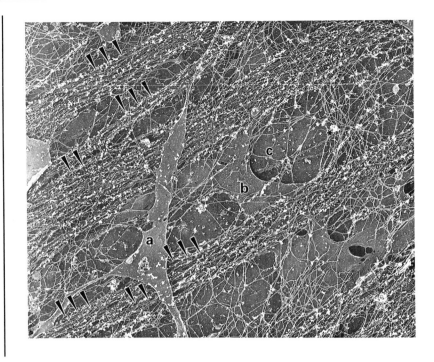

Quick-frozen fibroblast incubated with NaN_3 followed by antimyosin. The majority of actin filaments are aggregated into stress fibre bundles heavily labelled with variably spaced and sized non-aligned patches of antimyosin (arrowheads). Intermediate filaments (asterisk) course between these stress fibre bundles. Note the flaps of plasma membrane lying above (a) and below (b) the cytoskeleton. In places all the lower membrane has been removed by detergent to reveal the underlying glass coverslip (c). The bar represents 0.5 μm.

B

The actin filaments within a microvillus are highly ordered, and contribute to maintenance of the form of the microvillus. Actin filaments possess polarity, and have ends denoted *minus* and *plus* (the latter grow more rapidly and have a *barbed* appearance).

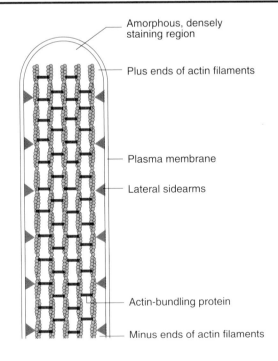

Amorphous, densely staining region

Plus ends of actin filaments

Plasma membrane

Lateral sidearms

Actin-bundling protein

Minus ends of actin filaments

An increasingly large number of proteins are being found to be associated with the actin filament system, with functions as indicated below.

Function of protein	Example of protein	Comparative shapes, sizes and molecular mass	Schematic of interaction with actin
Form filaments	Actin	50nm 370 × 43 kDa/μm	Minus end Plus end Preferred subunit addition
Strengthen filaments	Tropomyosin	2 x 35 kDa	
Bundle filaments	Fimbrin	68 kDa	10 nm
	α-Actinin	2 × 100 kDa	40 nm
Cross-link filaments into gel	Filamin	2 x 270 kDa	
Fragment filaments	Gelsolin	90 kDa	Ca²⁺
Slide filaments	Myosin	2 x 260 kDa	ATP
Move vesicles on filaments	Minimyosin	150 kDa	ATP
Cap the 'plus' ends of filaments and attach them to plasma membrane	(Not known)	?	
Attach sides of filaments to plasma membrane	Spectrin	α β 2 × 265 kDa plus 2 × 260 kDa β α	
Sequester actin monomers	Profilin	150 kDa	

A

Microtubules

Microtubules contribute another component of the cytoskeleton. They are formed from the protein *tubulin*.

The protein tubulin can polymerize and depolymerize for microtubules to form and dissociate. Tubulin consists of dimers, composed of α-tubulin and β-tubulin monomers. Microtubules are cylindrical polymers of tubulin, shown below (α-tubulin shown in pink, β-tubulin in white).

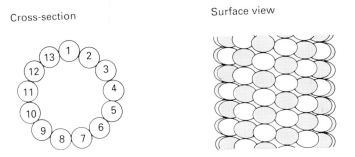

B

Polymerization of tubulin can be achieved in vitro, where it can be seen that an assembled microtubule consists of rows of tubulin molecules folded into *protofilaments*, about 25 nm in outer diameter. The faster-growing end is termed the *plus* end, the slower growing, the *minus* end. Polymerization requires GTP.

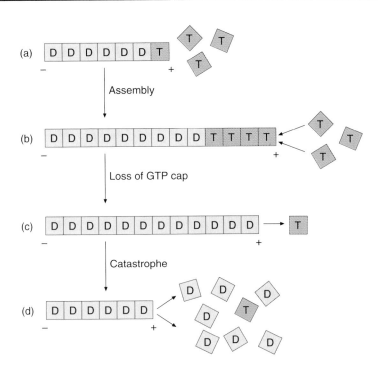

D = GDP
T = GTP

GTP is non-covalently bound to tubulin on both α- and β-subunits. At the β-site, GTP is hydrolysed to GDP more slowly than assembly of heterodimers at the plus end, to give a GTP *cap* at that end. GDP-tubulin is more readily disassembled than GTP-tubulin, so that the rate of GTP hydrolysis could regulate tubulin assembly. If the GTP cap becomes too small, rapid disassembly of tubulin (so-called 'catastrophe') can occur, with the possibility of 'rescue' if the concentration of heterodimers increases and GTP hydrolysis is slowed, a state known as *dynamic instability*. In the cell, after experimental disassembly of microtubules by the anti-mitotic compound *colchicine*, microtubules can be seen to grow from a small, perinuclear star-shaped stucture called an *aster*, this region being known as the microtubule-organizing centre (MTOC), the plus ends then growing from this point towards the cell periphery. In most cells the MTOC is the centrosome.

A

By using fluorescent antibodies against tubulin, microtubules can be visualized.

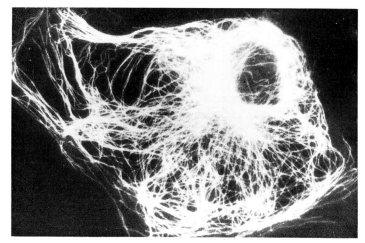

Microtubules visualized in a cell. The most dense staining surrounds the nucleus.

B

Intermediate filaments

Intermediate filaments contribute the third component of the cytoskeleton. These are formed from soluble forms of keratin, and other proteins such as *vimentin* and *desmin*. In the nucleus, lamins A, B and C are intermediate filament proteins that form the nuclear lamina and are important in maintaining the nuclear membrane (see p. 2C).

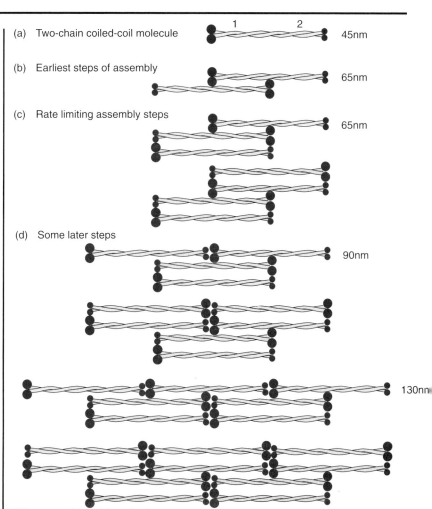

(a) Two-chain coiled-coil molecule   45nm

(b) Earliest steps of assembly   65nm

(c) Rate limiting assembly steps   65nm

(d) Some later steps   90nm

130nm

The assembly of keratin intermediate filaments is modelled in the scheme above. (a) A single coiled coil of one type I and one type II chain of keratin. (b–d) Successive steps in the assembly of larger aggregations.

A

Cell adhesion

The external surfaces of cells carry protein molecules that have the function of binding the cell to other cells or to substrates in the network of proteins and polysaccharides surrounding cells in tissues, known as the extracellular matrix. One of the most important of these cell surface receptors is one that binds *fibronectin*. The fibronectin receptor is one of a class of proteins termed *integrins*.

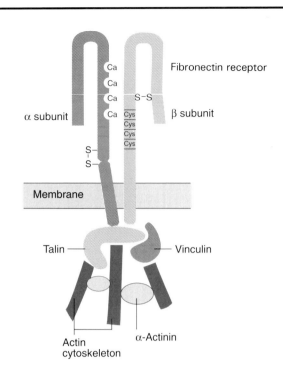

The fibronectin receptor has an α- and a β-subunit, the α-subunit being composed of two peptides joined by a disulphide bridge. On the inner surface of the plasma membrane, the fibronectin receptor makes contact with the actin filaments of the cytoskeleton through the proteins *talin* and *vinculin*. There is evidence of active communication with the cell interior by receptors such as the fibronectin receptor, to permit the cell to respond to extracellular events sensed by these receptors. The α- and β-subunits are members of families of homologous proteins, from which several families of integrins can be formed.

B   Fibronectin

The fibronectin receptor.

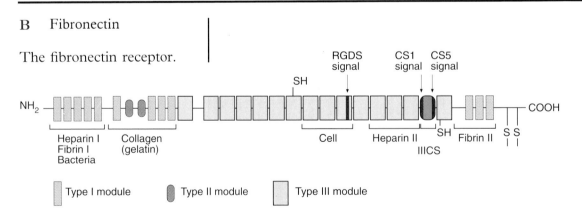

Fibronectin is a fibril-forming glycoprotein (of almost 2500 residues) found in the extracellular matrix. The domains of the molecule are composed of three types of homologous repeating unit, known as modules and named types I, II and III. Thus, five type I modules near the N-terminus form a heparin-binding domain. Other binding domains are indicated. Many integrins recognize the motif –RGDX– (where X is S, V, A, T, C or F). The so-called CS1 and CS5 signals in the IIICS region are cell adhesion signals with different specificities (CS1 towards lymphoid and certain tumour cells, CS5 towards melanoma cells).

A

Cell adhesion molecules are widely distributed on the surfaces of many cell types. These can be divided into two classes, typified by N-CAM (*neural cell adhesion molecule*) and E-cadherin (Ca^{2+}-dependent adhesion molecule).

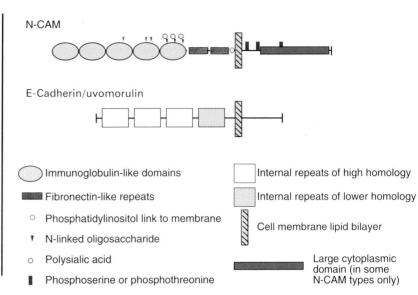

The key to the diagrams indicates the main features of these molecules. Both are capable of *homophilic* binding, that is, each can act as both ligand and receptor. The N-CAM family are Ca^{2+}-independent in contrast to the cadherins.

B

Lymphocytes and cell adhesion.

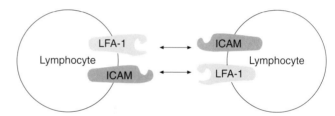

Lymphocytes carry the integrin LFA-1 (*lymphocyte function associated integrin*) which has an α_1 and a β_2 subunit. This can bind to the cell adhesion molecule ICAM (*intercellular adhesion molecule*). This type of interaction can occur in lymphocyte aggregation or adhesion to an antigen-presenting cell or target cell.

C

Integrins have been implicated in wound healing.

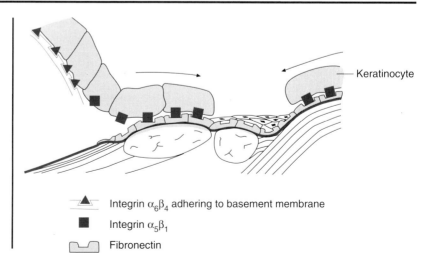

Integrin $\alpha_6\beta_4$ is a component of the hemidesmosome involved in binding keratinocytes to the basement membrane. On wounding, integrin $\alpha_5\beta_1$, which is a fibronectin receptor, is expressed, enabling the keratinocytes to migrate on fibronectin to cover the wound.

INDEX